The Epicurean Laboratory

Exploring the Science of Cooking

The Epicurean Laboratory

Exploring the Science of Cooking

Tina L. Seelig

Illustrated by Nancy J. Serpiello

W. H. Freeman and Company
New York

Library of Congress Cataloging-in-Publication Data

Seelig, Tina Lynn.
    The epicurean laboratory : exploring the science of cooking / by
Tina Seelig.
        p.   cm.
    Includes index.
    ISBN 0-7167-2099-X – ISBN 0-7167-2162-7 (pbk.)
    1. Cookery.   2. Food.   I. Title.
TX651.S423   1991
641.5–dc20   90-3855
                                        CIP

Printed in the United States of America

1 2 3 4 5 6 7 8 9 0 KP 9 9 8 7 6 5 4 3 2 1

For my parents,
Lorraine and Gerard Seelig

Table of Contents

Preface

❏ We often think of cooking as an art that our grandmothers taught our mothers and that French chefs keep hidden. Yet, cooking is also a fascinating science. In this book I hope to demonstrate many of the scientific principles involved in the art of cooking. Your new understanding will take away much of the mystery and uncertainty of cooking and provide you with new freedom to experiment in the kitchen.

❏ A chef in the kitchen is very much like a scientist in a laboratory. Both the chef and the scientist learn, using trial and error, which ingredients and conditions are necessary to create the desired results. By continually experimenting and keeping careful track of what has been done, the chef and the scientist make discoveries that are both exciting and rewarding. In addition, the tools that are used by a chef are surprisingly similar to those used by a scientist. The size and accuracy of the tools in a laboratory may be different from those of a kitchen, yet in the long run most do little more than mix, shake, separate, heat, cool, dissolve, and measure. More important, all the reactions that take place in the kitchen are founded on basic scientific principles.

❏ The goal of this book is to describe many of the scientific principles involved in cooking. The Introduction presents many of the scientific terms used in the book to initiate the nonscientist. Each chapter is followed by a recipe that uses some of the scientific concepts and ideas discussed in the chapter.

❏ I encourage you to read the descriptions of the scientific principles, to try the recipes, and to experiment with new recipes of your own. You will find that it is easy to be creative in the kitchen while developing a general understanding of science.

❏ If you are interested in reading more about the science discussed in this book there are many sources from which to choose. I recommend *Biochemistry* by Lubert Stryer, *Molecules* by P. W. Atkins, *On Food and Cooking* by Harold McGee, and *Introduction to Chemistry* by T. R. Dickson. In addition, if you have any comments for the author please feel free to contact me through the publisher, W. H. Freeman and Company, 41 Madison Avenue, New York, NY 10010.

❏ There are many people to thank for their help in preparing this book. First, I would like to thank Gary Carlson at W. H. Freeman and Company for his superb management of this project. Without his support this book would not have been possible. He sheparded the book through both calm and tortuous waters, making sure that the big picture was not lost. I would also like to thank my husband, Michael Tennefoss, for his unending enthusiasm about this project. He happily did all of the initial editing, gave me valuable advice, and encouraged me every inch of the way.

❏ I would like to thank Nancy Serpiello for designing and implementing the layout and graphics for the book. Her ability to turn standard scientific concepts into imaginative, three-dimensional images never ceased to amaze me. She was assisted by several people whom she would like to thank. Included are Mike Suh of W. H. Freeman and Company, who provided constructive comments and design suggestions; Steve Schoenberg and William Scott, who served as both creative design consultants and artists; and Chris Thomson, Laura Trang, and Laurie Carrigan, who assisted with the preparation of the artwork. Nancy's special acknowledgments go to Steve Schoenberg, whose patience and devotion were immeasurable; and to Steve Anaya for his encouragement of an apprentice many years ago.

❏ I would like to acknowledge both the work of Philip McCaffrey, who served as the project editor; and Julia De Rosa, who acted as production coordinator at W. H. Freeman and Company. I would also like to thank Diana Siemens for deftly editing the manuscript. In addition, I would like to thank Anne Ferril for professionally testing the recipes and for providing useful comments and suggestions.

❏ There were many scientists and food specialists who read various incarnations of the manuscript and provided invaluable comments on the scientific accuracy of the book. Included are Wayne Gisslen, Richard Kleyn, Carole McQuarrie, Christina Hammond, Mike Bellama, Alan Peterkofsky, Beverly Peterkofsky, Karl Drlica, Alan Lazarus, Ken Dill, Jolanda Schreurs, Hope Rugo, Martin Scott, and Steve Schoenberg. Each of these people provided useful suggestions that have shaped the final book. In addition, several lay readers provided comments on the readability of the text and the clarity of the presentation. They include Wayne Herkness, Wayne Wiener, Lucrecia Wiener, Christine Yemoto, and my parents, Lorraine and Gerard Seelig. Their recommentations were all very helpful.

❏ Finally, I would like to thank my wonderful son, Josh, for arriving on the scene in the midst of this project to pleasantly distract me from my work.

The Epicurean Laboratory

Exploring the Science of Cooking

Introduction

❏ Most of the chapters in this book require only an elementary knowledge of scientific principles and terminology. This introduction is designed for readers who have never studied chemistry and for those whose scientific vocabulary needs refreshing. It presents most of the important scientific principles and definitions on which the other chapters are based.

ATOMS AND MOLECULES

❏ All materials are composed of elements, such as carbon, hydrogen, oxygen, and nitrogen. An atom is the smallest unit of any element that is still identifiable as that element. Oxygen gas, for example, can be broken down into individual oxygen atoms, each of which has identical properties.

❏ All atoms contain a central nucleus surrounded by one or many shells of negatively charged (-) electrons. The electrons spin around the nucleus, much as the earth orbits the sun.

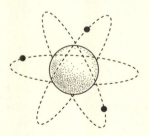

The nucleus is composed of two types of particles: positively charged (+) protons and uncharged neutrons. Because particles with opposite charges are attracted to each other, the negatively charged electrons within an atom are attracted to the positively charged protons within the nucleus. This attraction helps to hold the atom together.

❏ The atoms of different elements contain different numbers of electrons, protons, and neutrons. In most atoms, the number of electrons is equal to the number of protons. The positive and negative charges are, therefore, balanced and the atom is electrically neutral.

❏ An electron within one atom can be attracted to the protons of another atom. Sometimes, the electron will leave the first atom to join the second one. Because the donor atom is missing a negatively charged electron, it is positively charged. The receptor atom, in turn, gains an electron and acquires a negative charge. Both positively and negatively charged atoms are known as ions. The drawing following this paragraph depicts two atoms. As the electron from the left atom joins the right atom, the left atom becomes positively charged and the right atom becomes negatively charged.

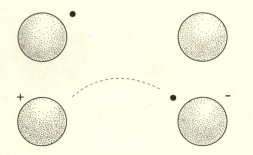

Some atoms share several electrons with other atoms and form more than one bond. In fact, an atom can form as many bonds with other atoms as it has electrons available to share. The drawing below shows a single atom sharing four pairs of electrons with four other atoms.

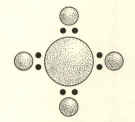

Table salt, or sodium chloride, for example, is formed when a sodium atom donates an electron to a chloride atom. The result is a positively charged sodium ion and a negatively charged chloride ion. These positively and negatively charged ions are attracted to one another and form an ionic bond.

❑ If an atom shares one electron pair with a neighboring atom, a single bond forms. If an atom shares two electron pairs with a neighboring atom, a double bond is created.

❑ When two atoms each have one electron to donate, the atoms may share the pair of electrons. When this happens, the two atoms form a covalent bond, as shown in the drawing below.

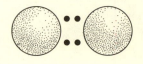

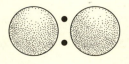

Therefore, an atom that has two electrons to share can form either two single bonds or one double bond. Double bonds are generally stronger than single bonds and hold the atoms together more tightly.

❏ A collection of two or more bonded atoms is called a molecule. Molecules come in innumerable shapes and sizes. Some molecules, such as water, contain only a few atoms; others, such as starch, contain hundreds. The properties of a molecule (including its melting point, color, and flavor) are determined by the number, type, and configuration of the atoms that make up the molecule.

❏ It takes energy to break the bonds that hold a molecule together. Energy, in the form of heat, causes the atoms within a molecules to vibrate. The higher the temperature, the faster the atoms vibrate. Eventually, the atoms vibrate so vigorously that the bonds between them break. Because double bonds are stronger than single bonds, more energy is required to break double bonds.

PROTEINS

❏ Amino acids are a type of molecule found in all living organisms. They are composed of only a handful of atoms, including carbon, hydrogen, oxygen, and nitrogen, and are the building blocks of all proteins, including hemoglobin, collagen, and insulin. There are 21 different amino acids, each with different properties. Amino acids link together in a long chain and then fold into a specific shape, forming a protein.

A protein's properties and functions are determined by the length of the amino acid chain, the specific amino acids in the chain, and the order of the amino acids.

❏ Each protein folds into a specific shape. These shapes resemble coils (helixes), flat sheets, doughnuts (channels), or balls (globules). The three-dimensional shape of the folded protein determines its characteristics and its function. The drawing below shows the shapes of several types of proteins.

❏ Some amino acids are attracted to water and are called hydrophilic (water loving). Other amino acids are repelled by water and are called hydrophobic (water fearing). When a protein folds, the hydrophobic amino acids cluster together inside the three dimensional structure of the protein, and the hydrophilic amino acids are arranged on the outside surface of the protein. As a result, the hydrophilic amino acids are exposed to the external, watery environment, and the hydrophobic amino acids are protected from exposure to water.

❏ Most proteins are held in their complex three-dimensional shape by the hydrophobic and hydrophilic interactions between the amino acids. Because these interactions are weak, proteins are easily unfolded by exposure to heat, acid, salt, or alcohol. When a protein unfolds, it loses most of its original properties and is said to be denatured. Only rarely can denatured proteins refold into their original shapes. The drawing below depicts the unfolding of a globular protein as it is denatured.

❏ Enzymes are an important class of proteins that act as catalysts. Catalysts are substances that speed up chemical reactions, but are not themselves changed by the reaction. Each specific enzyme can catalyze only one type of reaction. The illustration below shows how a molecule fits into a specific bonding site within an enzyme.

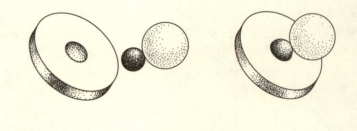

SUGARS

❏ Sugars, also known as carbohydrates, come in many sizes and perform different functions. Simple sugars, such as glucose and fructose, are made up of short chains that exist as single rings of carbon atoms to which oxygen and hydrogen atoms are attached.

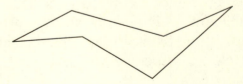

Complex sugars, such as starch, consist of many individual sugar molecules that are bonded together.

❏ Sugars serve as a short-term energy source for the cells that make up plants and animals. Cells are small, usually microscopic, fluid-filled sacs that contain a central nucleus and a surrounding membrane.

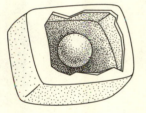

The nucleus contains DNA, a molecule that stores the genetic information for the cell; the fluid (or cytoplasm) within the cell contains the enzymes and other molecules necessary for the cell to survive; and the membrane is a selective barrier that allows some molecules to enter the cell and others to leave.

❏ When sugars are digested by cells, or are heated in the presence of oxygen, they are oxidized. Oxidation occurs when oxygen molecules interact with other molecules and strip away some of their electrons. This happens because oxygen has an extremely high affinity for electrons and pulls the electrons away from almost any molecule whose electrons are loosely bound.

❏ Oxidation of a molecule changes the molecule's appearance and properties. In addition, the energy stored in the molecular bonds is released. Many types of molecules can be oxidized. For example, when wood burns, oxygen in the air reacts with the wood molecules and the wood turns into ashes. In this reaction, energy is released in the form of heat.

SOLUTIONS

❏ Most chemical reactions, including those that occur in the kitchen, take place in solutions. A solution is composed of one substance, usually a liquid, in which one or more other substances are dissolved. Particles are considered to be dissolved when they are dispersed throughout the solution, creating a homogeneous mixture.

❏ An important property of solutions is their acidity. Acidity is a measure of the concentration of free protons (positively charged atomic particles) in the solution. Solutions with a high concentration of protons are acidic and those with a low concentration of protons are basic. A special scale called pH is used to measure the acidity or basicity of a solution. The pH of a solution is inversely related to the proton concentration. (This means that as the proton concentration increases, the pH decreases.) Pure water, which is neither acidic nor basic, has a pH of 7.0. A solution with a pH below 7.0 has a high proton concentration and is acidic. A solution with a pH above 7.0 has a low proton concentration and is basic. When acids and bases are mixed in equal amounts, they neutralize each other.

❏ The pH of a solution affects all the molecules in the solution. Proteins and sugars, for example, are very sensitive to pH. In an acidic solution, proteins are easily denatured, and complex sugars break down into smaller fragments.

SYMBOLS

❏ Symbols are used throughout this book to illustrate the molecules, bonds, cells, and forces discussed in the text. The symbols used in each chapter are de-signed to illustrate the specific concepts discussed in that chapter. In some cases, the symbols used reflect standard scientific notation; in others creative license has been taken.

❏ Different branches of science often use different symbols to represent the same chemical. Similarly, this book sometimes uses different symbols to represent the same molecule. For example, in Chapter 6, "Sugar," sugar molecules are drawn as cubes to illustrate how they stack to form a crystal, as seen in the drawing below.

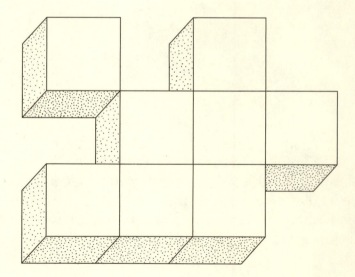

In Chapter 16,"Yeast," sugar molecules are drawn as stylized stars to show how the molecules break down when sugar is digested.

❑ Cooking takes advantage of many basic science principles that apply in the kitchen and throughout the universe. Knowing these principles will enable you to perform endless culinary experiments, and to view the world through the eyes of a scientist.

Proteins

1 Eggs

❏ Treating eggs in various ways changes their properties dramatically. Understanding the basic protein chemistry that underlies these changes allows you to control the reactions which, in turn, makes it easier to thicken custards, make meringues, and to leaven cakes and souffles.

❏ Eggs have an unusual property that gives them a central role in cooking: they are liquid at room temperature and solidify irreversibly as they are heated. This is in sharp contrast to sugar, fat, and water, which change reversibly from solids to liquids to gases as they are heated.

❏ The changes that take place in eggs when they are cooked are due to the proteins they contain. Eggs have many different types of protein, most of which are globular. Globular proteins are composed of chains of amino acids that naturally fold into a compact ball. Nonglobular proteins are made of amino acid chains that fold into other shapes, including sheets or coils, rather than balls. The illustration on this page represents three globular proteins.

❏ Heating an egg causes both globular and non-globular proteins to unfold. Unfolded proteins lose their original properties and are said to be denatured. As the proteins unfold, the amino acids that were once on the inside of the protein become exposed to the environment. They are then attracted to the amino acids of other denatured proteins and form weak bonds ∿∿∿ with them. This bonding process, called coagulation, results in a large three-dimensional network of proteins. The drawing on this page shows how three globular proteins unfold and then coagulate to form a network.

❏ If liquid ingredients such as milk are mixed with raw eggs, the liquid is held in the network of coagulated proteins that is created when the mixture is heated. The mixture starts out as a liquid and becomes progressively more solid as more bonds form. Overcooking creates too many bonds between the proteins, causing some of the liquid to be squeezed out of the protein network. This explains why, when overcooked, quiche becomes runny, custard gets lumpy, and scrambled eggs turn rubbery.

❑ Egg whites, which are composed of 10 percent protein and 90 percent water, coagulate at about 140°F (60°C). Egg yolks, which contain 15 percent protein, 35 percent fat, and 50 percent water, coagulate at about 154°F (68°C). A mixture of whites and yolks coagulates at an intermediate temperature.

❑ The coagulation temperature can be changed by adding other ingredients to the eggs. The addition of milk or water, for example, dilutes the egg proteins. Thus, it takes more heat to cause the proteins to move through the liquid to collide and bond. Milk or water, therefore, increase the coagulation temperature of eggs.

❑ Table salt or acids (lemon juice or cream of tartar) decrease the coagulation temperature of eggs. They do this by neutralizing the negatively charged portions of egg proteins. Once the egg proteins are neutralized, they have a greater tendency to bond together. Salt or acids may even cause eggs to coagulate without heat.

❑ Salt can also be used to coagulate proteins when hard boiling eggs. Adding salt to boiling water neutralizes and quickly coagulates any egg white proteins that leak out of cracks in the shell. The coagulated proteins seal the crack in the shell and prevent the egg white from continuing to stream out into the water, as shown in the drawing at right.

❏ Egg whites can also be denatured by whipping them with a beater. Whipping egg whites breaks many of the weak bonds between the globular proteins and also incorporates air into the mixture. The compact globular proteins unfold into a mass of long, tangled strands that surround the air bubbles. Some of these denatured proteins coagulate upon exposure to the air. This process produces a light, fluffy foam.

❏ Baking whipped egg whites causes many more of the proteins to coagulate, stabilizing the foam. In addition, the trapped air bubbles expand as the temperature rises, making the egg whites even fluffier. Because cream of tartar (an acid) promotes protein coagulation, it is often added to whipped egg whites to stabilize the foam. Liquid acids, such as lemon or lime juice, also coagulate proteins, but the added liquid inhibits the formation of foam. The drawing on this page shows the final structure of whipped egg whites.

❏ Some substances inhibit the formation of egg white foam by preventing the protein interactions that are necessary for bonding. Fats, for example, coat the denatured proteins and prevent them from coagulating. This is why it is important not to have any egg yolk (which contains fat) in an egg white mixture that you are planning to whip.

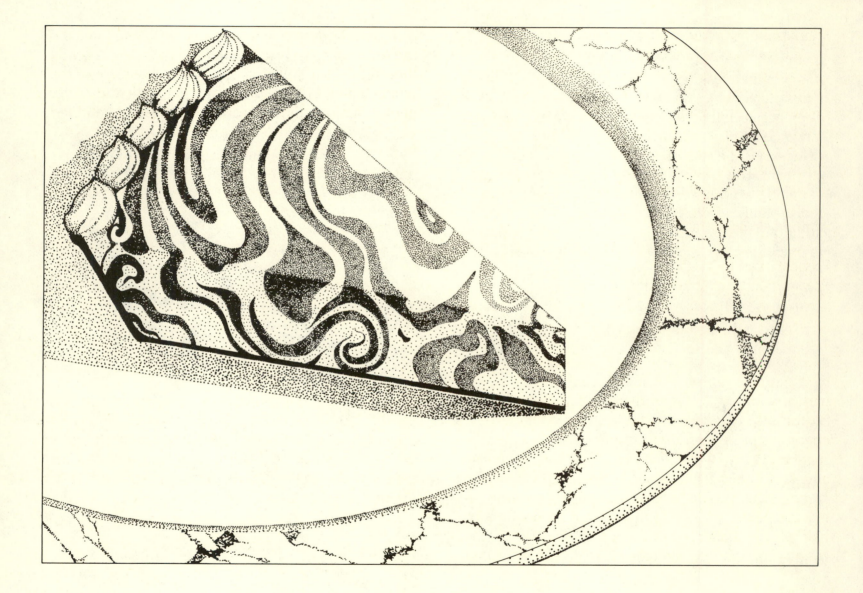

Chocolate Marble Cheesecake

5 tablespoons butter or margarine
1 1/2 cups chocolate wafer or graham cracker crumbs
3 8-ounce packages of cream cheese
1 1/4 cups sugar
16 ounces sour cream
4 large eggs
2 teaspoons vanilla extract
4 or 5 teaspoons lemon juice
1/3 cup semisweet chocolate chips
1/2 cup vanilla yogurt
Almonds for decoration

Preheat oven to 350°F
Preparation time: 2 hours
Yield: 8 to 10 servings

1. In a saucepan, melt the butter or margarine and thoroughly mix in the cookie crumbs. Press the mixture onto the bottom of an 8- or 9-inch springform pan and set aside or refrigerate. 2. In a heavy-bottomed saucepan over low heat, melt the cream cheese, stirring constantly, until smooth. Add the sugar and blend well. 3. Transfer the cream cheese and sugar mixture to a large bowl. Add the sour cream, eggs, vanilla, and lemon juice. Mix with an electric beater or by hand until well blended. Pour half of the batter over the crust mixture in the springform pan. 4. Melt the chocolate in the top of a double boiler or in a microwave oven. Pour the melted chocolate into the remaining batter and mix thoroughly. 5. Pour the chocolate batter over the white batter in the springform pan. Stir the batter gently with a large spoon to create a marbled effect. 6. Bake in a 350°F preheated oven about 1 1/2 hours, until the cake has risen about 2/3 of the way from the outside edge. *The cake is removed from the oven before it is actually done because it continues to cook. If it cooks too long, the egg proteins will form too many bonds and squeeze out the enclosed liquid, creating a dry, grainy texture.* 7. Remove the cake from the oven and allow it to cool to room temperature. 8. Remove the outside rim of the springform pan and spread the top of the cake with vanilla yogurt if desired. Decorate with almonds. Refrigerate the cake to cool it completely before serving.

Variations:
1. Substitute 3 tablespoons amaretto for the vanilla. 2. Instead of adding the melted semisweet chocolate chips to half of the batter, pour all of the white batter into the springform pan and top the cooled cake with yogurt and fresh strawberries, cherries or kiwifruit.

2 *Milk*

❏ Milk is produced by the mammary glands of female mammals and is the sole source of nutrition for their rapidly growing young. Humans are the only animals who consume the milk of other mammals and who continue to drink it after infancy.

❏ We most often drink cow's milk, which is similar to the milk of other mammals. It is composed of water, fat, protein, and lactose (milk sugar), as shown in the drawing on this page. To understand the properties of milk, it is helpful to know the characteristics of each of its components.

❏ Milk fat is contained in globules that are suspended in water. Since fat is lighter than water, these globules would be expected to rise to the surface. They are inhibited from doing so by natural emulsifiers within the milk. Emulsifiers are molecules that have two ends, one of which dissolves in fat and the other of which dissolves in water. The fat-soluble end immerses itself in a fat globule, while the water soluble end extends out into the watery surroundings. A fat globule that is completely surrounded by emulsifiers is like a pincushion stuck with pins. It is much less likely to pool with other fat globules and to float to the surface. (Several types of emulsions are described in Chapter 15, "Oil and Water.")

❏ If fresh milk is allowed to stand, the fat globules ultimately float to the top and form a layer of cream. Most milk that we drink today is homogenized, meaning that the fat globules are broken into very tiny spheres by pumping the milk through a valve at high pressure. Because these tiny spheres of fat cannot pool together easily, they do not float to the surface. (Milk is also pasteurized, a process that uses heat to destroy disease-causing microbes, ensuring the safety of the milk.)

❏ It is the fat in milk that makes it feel thick and rich in your mouth. Lowfat and nonfat milk seem much thinner. The fat also makes the milk opaque because the fat globules reflect light. When fat is removed, milk appears translucent because light can more easily pass through the milk.

❏ Milk protein is categorized into two major groups: casein and whey proteins. The larger group, the casein proteins, exist in small, negatively charged clusters that gently repel one another. Each cluster is composed of numerous individual protein units. These clusters are similar to tiny sponges, as they contain approximately seventy percent water. The drawing on this page illustrates a casein protein cluster.

□ Acids, salt, or very high heat cause casein protein clusters to lose their charge. Once the charge is gone, the clusters lose their mutual repulsion and bond www to one another, or coagulate, forming large curds in the milk. These curds, as shown in the illustration on this page, enclose an enormous volume of water. Limiting the salt, heat, and acidic ingredients used in recipes containing milk prevents the formation of curds.

□ Whey proteins also coagulate when heated, but they do not coagulate when exposed to acid or salt. These proteins stay in the water, also called the whey, when the casein proteins coagulate. (See Chapter 19, "Yogurt," for additional information on curds and whey.)

❏ Cooking has a major effect on both types of milk proteins. When milk is heated in an open pot, a thick skin forms on the surface. This layer is created when water evaporates from the surface of the milk, leaving a larger local concentration of casein and whey proteins. The high concentration of proteins has a greater tendency to coagulation and, therefore, forms a skin. If the skin is removed during cooking, a new skin forms as more water evaporates. One way to prevent skin formation is to cover the pot to keep the surface of the milk from drying out.

❏ Milk will also burn or scorch if it is heated too quickly. Heat denatures some of the whey and casein proteins, causing them to become less soluble in water. They settle on the bottom of the pot and the direct heat under the pot causes them to burn. Scorching of milk can be prevented by using a heavy pot that more efficiently disperses the heat, by using a double boiler to keep the temperature below the boiling point of water, or by constantly stirring the milk to prevent the proteins from settling on the bottom of the pot.

❏ Lactose is a type of sugar found only in milk. It is composed of two smaller sugar molecules, glucose and galactose, that are bonded together. Lactose is responsible for making milk taste somewhat sweet. A special enzyme called lactase is needed to break lactose into its component sugars. In some people, this enzyme is present only for the first few years of life, when a child is most likely to drink its mother's milk. When children or adults who lack this enzyme consume milk, they are unable to digest the lactose. When the undigested lactose reaches the large intestine (colon), the bacteria normally living there consume the lactose and produce carbon dioxide gas. Lactose-intolerant people usually decide not to consume milk products because the resulting gas causes discomfort.

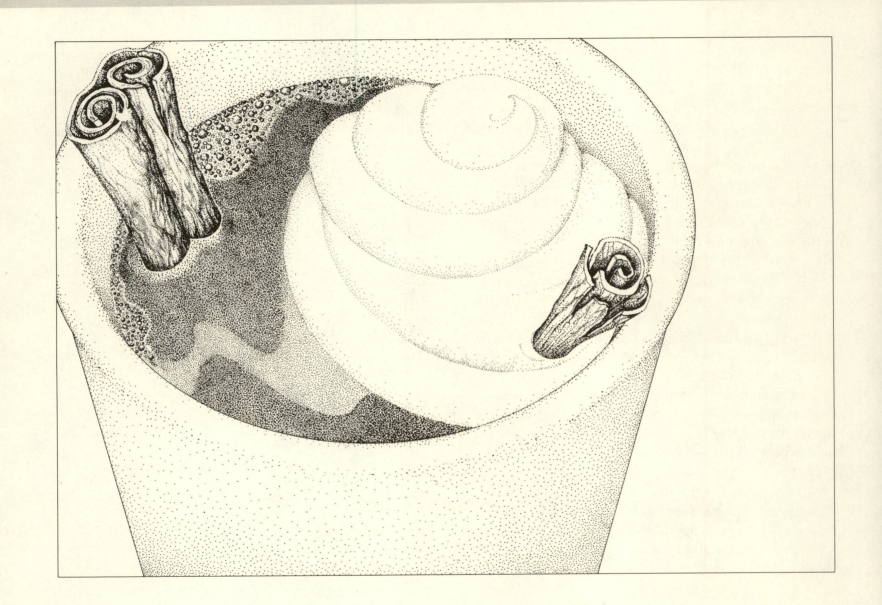

Mexican Hot Chocolate

2 cups milk
2 heaping teaspoons sweetened cocoa powder
    or 1$^1/_2$ teaspoons unsweetened cocoa and sugar
    to taste
$^1/_4$ teaspoon ground cinnamon
$^1/_2$ teaspoon vanilla extract
Whipped cream and ground cinnamon (optional)

Preparation time: 5 minutes
Yield: 2 servings

1. In a heavy-bottomed saucepan or double boiler, heat the milk slowly until it is steaming, stirring constantly. *The heavy saucepan helps disperse the heat evenly, and the double boiler prevents the temperature from exceeding the boiling temperature of water. These methods prevent the milk from scorching.* **2.** Stir in the cinnamon, cocoa, and vanilla. **3.** Pour the mixture into warm mugs. Top with whipped cream and ground cinnamon if desired. Serve immediately.

3 Meat

❏ All meat has a complex cellular structure, composed of many types of proteins, fats, water, and carbohydrates. Each of these components is affected by the heat of cooking.

❏ Most of the meat that we consume is derived from the skeletal muscles of animals. There are two major types of muscle cells: red fibers and white fibers. Whether meat is light- or dark-colored is determined by how much of each cell type is present. Light meat has a higher proportion of white fibers and dark meat has a higher proportion of red fibers. The two muscle cell types have different structures, functions, textures, flavors, and reactions to cooking.

❏ Red muscle fibers, also known as slow-twitch fibers, must sustain long periods of activity in the animal. To maintain this activity, fat ⬤ is stored in and around these muscles as an energy source. Oxygen ⬤⬤ is necessary to metabolize the fat. A specialized molecule called myoglobin ⬤⬤ stores the needed oxygen. Like hemoglobin in the blood, myoglobin is dark red when oxygen is bound to it.

❏ Dark meat is composed of a high percentage of red muscle fibers and, therefore, contains large stores of fat and myoglobin. Chicken legs, for example, contain a high percentage of red muscle fibers because these muscles are active for extended periods. A stylized drawing of red muscle fibers is shown below.

Cooking changes the color of dark meat. A small amount of heat causes the myoglobin to take up oxygen, so the meat stays red. More heat causes the myoglobin to release oxygen, turning the meat purplish. Continued heating causes oxidation of the iron contained in the myoglobin, causing the meat to turn brown. This is why rare meat appears red and well-done meat appears brown.

White muscle fibers, also called fast-twitch fibers, are found in muscles that are very active for short periods of time. These cells use sugar circulating in the blood as a source of energy. Since white muscle fibers can metabolize sugar with or without oxygen, myoglobin is unnecessary as a source of oxygen. Thus, white meat has a light color and little fat. Chicken breasts, which the chicken uses only for short bursts of activity, are an example of meat that is composed primarily of white muscle fibers. The drawing above depicts white muscle fibers.

❏ Meat that contains a large percentage of fat is very tender. When meat is heated, the fat melts and lubricates the muscle fibers. This allows the fibers to slide past one another easily when the meat is chewed.

❏ A meat's texture is affected by the type as well as the amount of fat present within it. Saturated fats pack together tightly and are solid at room temperature. Meats that contain high percentages of saturated fat, such as beef or pork, have firm textures. Unsaturated fats, on the other hand, do not pack together and are liquid at room temperature. Chicken and fish, which contain high percentages of unsaturated fat are soft textured. (See Chapter 14, "Fat & Oil," for a more extensive description of saturated and unsaturated fats.)

❏ In addition to fat, meat also contains many types of protein. A predominant one is collagen, which makes up the connective tissue of the muscle. The amount of collagen in the meat affects its texture: the more collagen, the tougher the meat. Fish, which has little collagen, is very tender while beef, with a much greater collagen content, is tougher. Even within the same animal, the amount of connective tissue varies from muscle to muscle. This is why some cuts of meat are tougher than others.

❏ Meat can be tenderized in several ways before it is cooked by denaturing some of the collagen. Marinating meat in acid, such as wine or vinegar, denatures some of the collagen on the meat's surface. Natural enzymes can also be used to break the collagen into smaller pieces. These enzymes include bromelain from raw pineapple, ficin from figs, and papain from papaya. The enzymes can be used to tenderize meat by adding one of these fruits to a marinade or by applying a commercial extract of the enzymes to the meat. Meat can also be tenderized by grinding it into small pieces, as with hamburger, which makes it easier to chew. This drawing depicts various tenderizing methods.

Cooking can also be used to tenderize meat. For example, boiling or stewing meat denatures the collagen and eventually turns it into gelatin, which is much softer than collagen. This is why stewed meats literally fall apart. (See Chapter 4, "Gelatin," for a further description of the transformation of collagen into gelatin.)

Many structural changes take place when meat is cooked. First, the muscle cells break open and spill their contents, which then mix with the contents of other cells. Continued heat denatures the proteins, exposing amino acids that were once on the interior of the proteins. Amino acids on adjacent denatured proteins then bond together or aggregate. The aggregated proteins form a network that traps the fluid components of the meat. As the meat continues to cook, more bonds form between the amino acids, and the resulting network contracts. As the network tightens, fluid is squeezed out of the meat. This is why well-done meat is drier than rare meat. The drawing on this page shows the structure of cooked meat.

❏ Cooked meat is much more solid than raw meat because most of the fluid in cooked meat is trapped within the protein network. However, when it is cut, cooked meat leaks juice. This happens because the trapped water easily leaks out of the damaged protein network, as shown in the drawing on this page.

❏ There are many ways to cook meat, including broiling, frying, roasting, barbecuing, steaming, boiling, and microwaving. These methods vary in the amount of heat used, the proximity of the meat to the heat source, and the length of time that the meat is cooked. Cooking meat in a stew, for example, takes a long time because the temperature of the meat never rises above the boiling point of the stew broth. Frying, on the other hand, cooks the meat very quickly because it employs direct heat at a high temperature. In all traditional cooking methods, heat is applied to the outside of the meat, which causes the surface to cook first. As the outside of the meat becomes hot, the heat is conducted inward, and the center is then cooked. Alternatively, when meat is cooked using a microwave oven, the water throughout the meat is heated to its boiling temperature. Cooking meat in a microwave oven, therefore, produces many of the same effects as boiling.

❏ Different cooking methods produce different flavors and textures. A flavorful brown crust, for example, is created by direct high heat. The browning is caused

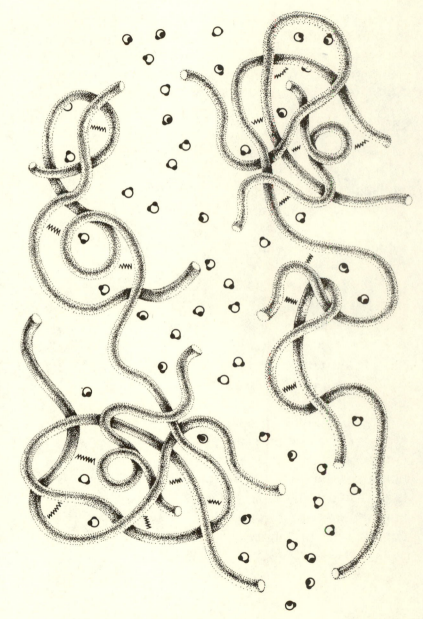

by chemical reactions between the carbohydrates and the amino acids on the surface of the meat. These reactions can take place only at very high temperatures. When meat is boiled or stewed, browning cannot take place because the temperature never gets high enough for these reactions to occur. This is why some recipes suggest browning meat first to develop color and flavor, and then stewing it to make the meat tender.

❑ In summary, cooking has many effects on the complex structure of meat. It causes the myoglobin to lose its oxygen, the fat to melt, the connective tissue to become partially denatured, and proteins to form a network enclosing most of the water. The texture and flavor of cooked meat are determined by the type of muscle fibers, the amount and type of fat, the quantity of connective tissue, any tenderizing that has been employed, and, finally, how the meat is cooked.

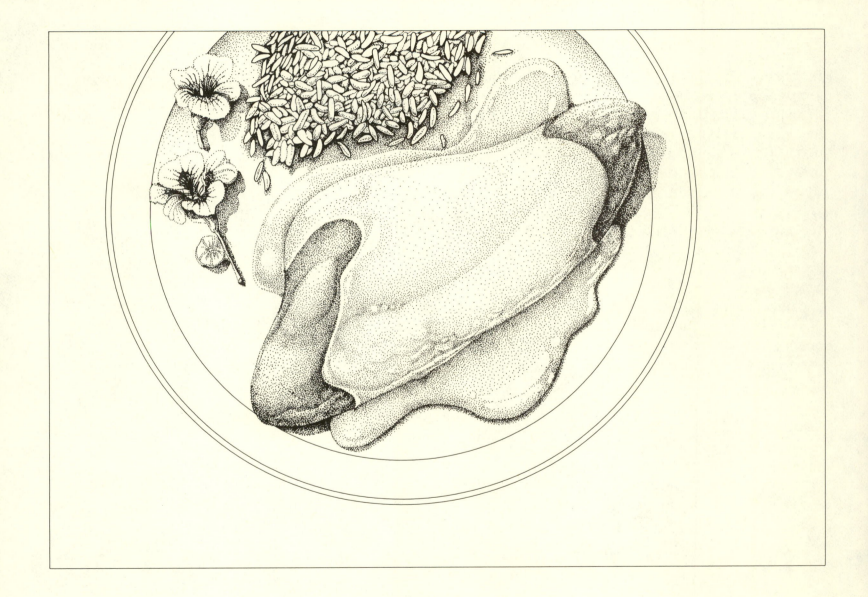

Honey Mustard Chicken

7 or 8 boneless half chicken breasts, skin removed
1/3 cup honey
1/3 cup Dijon mustard
1 teaspoon curry powder
1/2 teaspoon ground ginger

Preheat oven to broil
Preparation time: 30 minutes
Yield: 6 to 8 servings

1. Rinse the chicken breasts in cold water, pat dry, and put them in a baking dish. **2.** In a separate bowl, combine the honey, mustard, curry powder, and ginger. Spread the mixture over the chicken. **3.** Broil the chicken 4 to 6 inches from the flame for about 5 minutes or until the honey-mustard mixture is bubbly and slightly brown. **4.** Bake the chicken at 350°F for 15 to 20 minutes until the meat is white and it leaks clear juice when cut with a knife. *Juice leaks from the meat because the protein network that forms when the meat is heated contains most of the fluid components of the meat.* **5.** Serve immediately with white or brown rice and steamed vegetables.

Variations:
1. Add 1 or 2 cloves of crushed garlic for a stronger tasting sauce. **2.** Add 3 or 4 tablespoons of plain yogurt for a milder sauce. **3.** Marinate the chicken in the sauce for a few hours to make it more flavorful.

4 Gelatin

❏ When you dissolve gelatin in water and then cool the mixture in the refrigerator, it thickens and becomes firm. This behavior is explained by the properties of the proteins that make up gelatin.

❏ Gelatin is actually denatured collagen which, as discussed in Chapter 3, "Meat," is a protein that is present in the muscle tissue of animals. It is also found throughout the bodies of most animals, including the skin, tendons, bones, and the cornea of the eye.

❏ Collagen, as shown in the drawing, is a triple helix. It is composed of three separate chains of amino acids, each spiraling around the others. The chains are held together in the helical structure by many weak bonds ∿∿∿.

When collagen is heated in water, most of these weak bonds break, and the protein is denatured. It then resembles an unbraided length of rope and is called gelatin. The drawing on this page shows what happens to the structure of collagen when it is denatured.
❏ To use prepackaged, dry gelatin in cooking, it is best to disperse it in cold water and then to heat the water to dissolve the gelatin. Adding gelatin directly to hot water causes lumps to form because the hot water is absorbed before the gelatin is completely dispersed in the water. Cold water is not absorbed by the gelatin as quickly as hot water, and therefore allows the gelatin to disperse evenly without forming lumps. Because gelatin may also be dispersed in sugar, presweetened dry gelatin can be dissolved directly in hot water. (A similar phenomenon is discussed in Chapter 7, "Starch.")

❑ When dissolved gelatin is added to other liquids and then refrigerated, a semisolid gel forms. The gel is created when the weak bonds that originally held the collagen in its three-dimensional shape begin to form again. These new bonds are completely random, unlike the highly ordered bonds in the original collagen. In fact, once broken, the original, triple helix structure will never form again.

❑ The bonded proteins in the gel create a large network that traps the liquid ingredients. Continued cooling causes more bonds to form, and the gel becomes more rigid. The drawing on this page shows the structure of a gel.

❏ Several factors affect the formation of the final gel. One factor is the relative proportion of gelatin to the liquid ingredients. Only a very small amount of gelatin (1 part gelatin to 99 parts water, by weight) is capable of immobilizing a liquid. The more gelatin in the mixture, however, the firmer the gel.

❏ The other ingredients in the recipe also influence the gel. For example, the acidity and the amount of sugar greatly influence the firmness of the gel. A gel forms best when the ingredients are slightly acidic ($pH = 5$) and when a small amount of sugar is included in the recipe. Under these conditions, the denatured collagen molecules are best able to interact and bond together.

❏ A gel will not form at all if fresh pineapple is present in the mixture. Fresh pineapple contains the enzyme bromelain, which breaks the protein chains into small fragments that cannot gel. (This enzyme is used to denature collagen when tenderizing meat. See Chapter 3, "Meat," for details.) However, cooked or canned pineapple will not prevent gelling, because bromelain is destroyed by heat.

❏ Gelatin is used in many recipes, including aspic, pie, and mousse, to create a semisolid consistency. Gelatin is not the only substance that can change the texture and structure of a solution, however. Eggs and starch, as described in Chapters 1 and 7, respectively, are also used to thicken various foods.

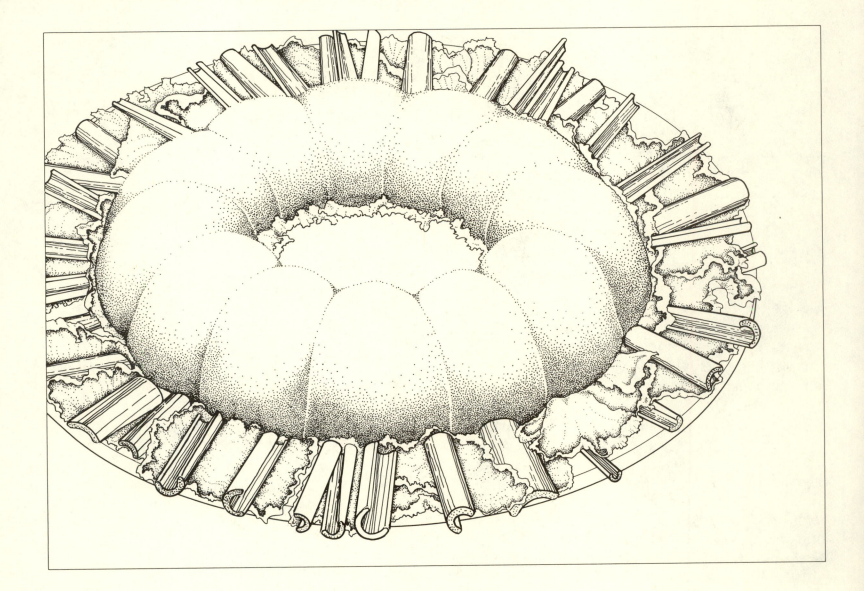

34

Salmon Mousse

1 pound fresh (cooked) or canned boneless salmon
$\frac{1}{2}$ cup diced celery (2 large stalks, leaves removed)
$\frac{1}{2}$ cup diced green pepper ($\frac{2}{3}$ of a medium-size
  pepper)
$\frac{1}{4}$ cup finely chopped onion ($\frac{3}{4}$ of a small onion)
3 or 4 tablespoons chopped fresh dill
1 cup mayonnaise
1 8-ounce package cream cheese
1 can concentrated tomato soup
$\frac{1}{2}$ cup cold water
2 teaspoons (1 envelope) unflavored, unsweetened
  gelatin

Grease a 6-cup mold
Preparation time: $6\frac{1}{2}$ hours, including chilling time
Yield: 12 servings

1. In a large bowl, break salmon into small pieces and remove any bones. Add the celery, green pepper, onion, dill, mayonnaise and mix thoroughly. 2. In a heavy-bottomed saucepan, melt the cream cheese. Add the tomato soup and continue cooking over low heat, stirring constantly, until the mixture is smooth and creamy. Pour the cream cheese and tomato soup mixture into the salmon mixture and blend thoroughly. 3. Pour the cold water into a small pot and stir in the gelatin. *Cold water is needed to disperse and soften the gelatin before it is heated.* 4. Slowly heat the water and gelatin just until the gelatin dissolves. *Do not boil the gelatin because boiling can denature the protein to such an extent that it cannot form a gel.* 5. Add the dissolved gelatin to the salmon mixture and blend well. Pour the mixture into the greased mold and refrigerate for 6 hours, or until it is firm. 6. Unmold by placing the bottom of the mold in hot water for a few seconds and turn the mold over onto a large serving dish. Refrigerate until ready to serve. 7. Serve with celery, carrot sticks and pumpernickel bread.

Variations:
1. Add a few dashes of tabasco sauce to give the mousse a bite. 2. Substitute plain yogurt for the mayonnaise to make a lighter mousse. 3. Add 3 tablespoons of lemon juice to the mousse.

5 Garlic & Onion

❏ Through the ages, garlic has been promoted as an aphrodisiac, a vampire repellent, a builder of strength, a cure for high blood pressure, and a poison. Much of this folklore is based on the strong smell that garlic produces when it is cut open. What causes this smell?

❏ Many plants, including garlic, produce small quantities of odiferous compounds as a defense mechanism against bacteria, insects, and some animals. It is ironic that many people have developed a special taste for the same qualities of garlic that evolved to protect the plant from being eaten.

❏ Garlic cells contain an odorless molecule called alliin . As shown in the illustration on this page, alliin is contained inside cells, while an enzyme, alliinase, is stored in the intercellular spaces (the spaces between individual cells).

❏ Enzymes are specialized proteins that catalyze reactions. They are shaped in such a way that they can hold other molecules in a fixed position. Once a molecule is held by an enzyme, it is more likely to undergo specific chemical reactions.

❏ Cutting a bulb of garlic disrupts the cell membranes and allows the alliin to come in contact with the alliinase. The alliin molecule fits into the alliinase molecule as a key fits into a lock. When these molecules interlock, a complex series of reactions is triggered. The alliin molecule is broken into several new molecules with different properties.
As with all enzymes, each alliinase molecule catalyzes many reactions in a short period of time without itself changing. The drawing below shows how the alliin fits into the alliinase once the garlic cells have been damaged.

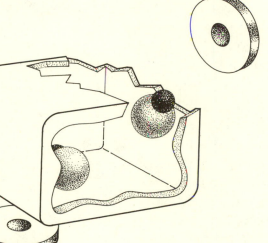

❏ One of the molecules produced by the enzymatic reactions is the primary source of garlic's odor. As more cellular membranes are disrupted, more reactions take place, more of the molecules (including the odiferous ones) are produced, and the garlic smell increases. The odiferous molecules ultimately dissipate in the air or are broken down by cooking.
❏ It is easy to minimize garlic's pungent odor. Since the enzymatic reactions take place only when the garlic cells are cut open, the smell can be reduced by chopping the garlic less finely. Cooking garlic also breaks down the offending molecules. In fact, if you bake or boil a whole garlic bulb, the alliin molecules are transformed into new, larger molecules before they are able to react with the alliinase. The newly created molecules have a pleasant, sweet taste.
❏ The onion is closely related to garlic. Like garlic, it causes no distress until it is cut, whereupon it makes people cry. What happens when an onion is cut?
❏ As with garlic, there are molecules within the onion cells that are inactive until they come in contact with enzymes located outside the cells. Cutting an onion ruptures the cellular membranes and allows the molecules to react with the enzymes. The new molecules that are formed are similar to those produced by garlic. One of these

molecules (called a lachrymator, or tear-producing compound) becomes airborne and reacts with fluids in the eye, forming sulfuric acid. This is why the mist from a freshly cut onion burns our eyes.

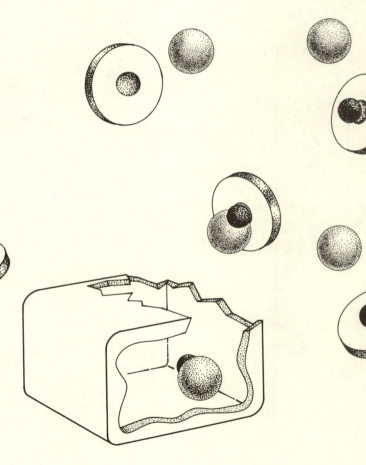

❑ Luckily, the lachrymator from onions is very unstable and quickly decomposes into a less annoying
molecule. This decomposition is hastened by cooking. In addition, the lachrymator is water soluble and
is rendered harmless if the onion is cut under water.
❑ The drawing on the left-hand page shows the new
molecules formed by the enzymatic reactions in both
onion and garlic. The drawing on the right-hand page
illustrates how these new molecules float away into
the air and ultimately reach our noses and eyes.
❑ Both onions and garlic are quite mild until their
cells are damaged. Rupturing the cell
membranes triggers a series of reactions in which the molecules
that were inside the cells react
with enzymes stored outside
the cells. The newly formed
molecules are intended by
nature to keep predators
away. Although this reac-
tion evolved to protect these
plants, we have learned to
enjoy their "poisons" and to
control them for our pleasure.

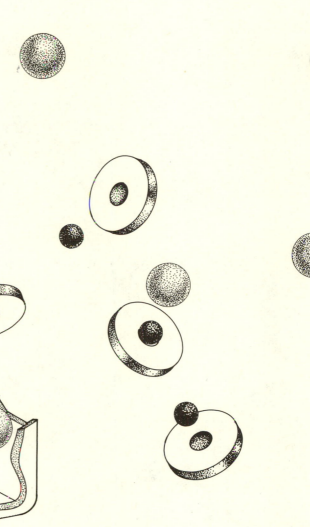

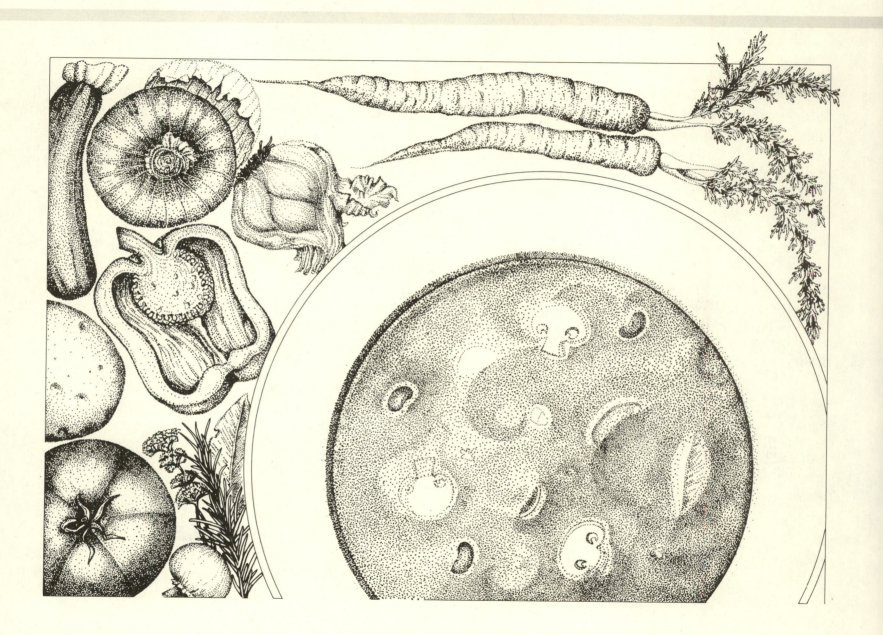

Spicy Vegetable Soup

3 tablespoons olive oil
1 large onion, coarsely chopped
2 large garlic cloves, crushed
4 medium-size carrots, peeled, trimmed, and sliced
2 cups quartered mushrooms (about 8 ounces)
2 medium-size zucchini, sliced and quartered
2 medium-size potatoes, cut in ½ inch cubes
3 large tomatoes, coarsely chopped
1 15-ounce can red kidney beans, drained
6 cups water or chicken stock
1 10-ounce can tomato sauce
½ cup dry red wine
2 teaspoons dried basil
3 tablespoons chopped fresh parsley (a small bunch)
3 dried bay leaves
1 teaspoon dried rosemary
¼ teaspoon red pepper
Salt and pepper, to taste

Preparation time: 1¼ hours
Yield: 6 to 8 servings

1. Heat the olive oil in a heavy-bottomed stockpot. Add the onion and garlic and saute for 5 minutes. *Cooking the onion and garlic causes the strong odor-producing molecules to change into mild, sweet-tasting molecules.* **2.** Add the carrots and sauté for another 5 minutes. **3.** Add the mushrooms, zucchini, and potatoes and sauté for another 5 minutes. **4.** Add the tomatoes, kidney beans, and the water or chicken stock and mix. **5.** Add the tomato sauce, red wine, and herbs. Simmer for at least 30 minutes until all the vegetables are tender and the flavors have blended. **6.** Serve with a sprinkling of grated Parmesan cheese along with fresh bread and red wine.

Variations:
1. Add pasta shells 15 minutes before the soup is served. **2.** Remove several cups of the soup and purée in a blender. Add the purée to the simmering soup to thicken it. **3.** Use any other combination of vegetables you choose.

Carbohydrates

6 Sugar

❏ Many people feel a strong urge to eat sugar. This urge was probably very adaptive for early humans because most fruits and vegetables that have a high sugar content are ripe and are not poisonous. People have since developed methods for producing refined sugar from sugar beets and sugar cane and have found many ways to use sugar in cooking.

❏ Sugar is used as a sweetener in cakes, sauces, preserves, custards, salad dressings, and many other foods. It is also used as a source of food for yeast in bread dough. (See chapter 16, "Yeast.") This chapter will focus on the properties of sugar when used in highly concentrated solutions and will explain how sugar is used to make candy.

❏ Sugars are composed of relatively simple molecules. Some sugars, such as fructose and glucose, are individual molecules ▱ ▱ , while others are composed of several molecules. A sucrose (table sugar) molecule, for example, is composed of one glucose molecule and one fructose molecule that are bonded together.

❏ Identical sugar molecules tend to stack together in a regular pattern and form a solid crystal. For example, a collection of identical sucrose molecules is solid at room temperature. The resulting crystal reflects the structure of the individual molecules. The drawing on this page illustrates individual sugar molecules, represented as single cubes, and depicts the crystal that forms when these molecules stack together tightly.

Dissimilar sugar molecules interfere with crystallization by disrupting the repetitive pattern of identical sugar molecules. Thus a mixture of more than one type of sugar tends to be liquid at room temperature. Honey, maple syrup, and corn syrup are mixtures of different types of sugars; these mixtures remain liquid at room temperature. The drawing below depicts a collection of nonidentical sugar molecules.

Heating crystalline sugar causes it to melt into a clear liquid because energy from the heat disrupts the orderly crystalline structure. Further heating of the liquid breaks the individual sugar molecules into smaller fragments. Some of these fragments evaporate, producing the pleasant smell of caramelized sugar. Other molecules are darkly colored and become trapped in the mixture, creating a rich brown liquid. If the mixture is heated too much, the molecules eventually burn.

Crystalline sugar will dissolve in water. Once the water is saturated with sugar, however, no more will dissolve. Any sugar added beyond the saturation point will precipitate (fall to the bottom as crystals). Heating a saturated sugar solution increases the amount of sugar that can be dissolved. This happens because heating a liquid causes the molecules to move faster. The rapidly moving water molecules literally can suspend more dissolved particles.

The hotter a solution is, the more particles will dissolve in it. If a hot saturated sugar solution is cooled, then more sugar remains dissolved than normally would be at the cooler temperature. The solution is said to be supersaturated. The cooler the solution gets, the more supersaturated it becomes. Supersaturated solutions are very unstable. If such a solution is disturbed even slightly, by stirring the mixture, for example, the dissolved sugar will recrystallize.

❑ When a sugar solution is boiled, water evaporates and the sugar concentration increases. The higher the concentration of dissolved sugar, the higher the boiling temperature. (See Chapter 10, "Water," for a detailed discussion of how boiling temperature changes with increasing concentrations of dissolved solids.) In fact, the concentration of sugar in a solution can be determined indirectly by measuring the boiling temperature of the mixture. This is why we use a candy thermometer when making candy.

❑ Water without any dissolved solids boils at 212°F (100°C). When a solution is approximately 70 percent sugar, the boiling temperature is about 230°F (111°C), and a small amount of the solution will become threadlike when dropped into cold water. If a solution is boiled further, evaporating more water, the boiling temperature will increase to about 240°F (116°C) at a sugar concentration of about 80 percent. At this point, a small amount of the solution will form a soft ball when dropped into cold water. At a sugar concentration of 90 percent, the solution boils at about 255°F (125°C) and forms a hard ball when dropped into cold water. When the solution reaches 98 to 99 percent sugar, it boils at about 300° F (150° C) and forms a hard ball that cracks in cold water. Thus, by using either a candy thermometer or the "thread, ball, crack" method, it is possible to determine the sugar concentration of a solution.

❑ When we make candy, we create a supersaturated sugar solution and carefully control the cooling process. To make candies such as sour balls, brittles, and lollipops, the sugar solution is heated to the hard crack stage (only about 1 to 2 percent water) and then cooled very quickly by pouring it directly onto a cool surface. The solution may also be cooled rapidly by forming it into small balls that lose heat rapidly. These candies are amorphous or noncrystalline, that is they have no repetitive structure, because the solution is not allowed to crystallize as it cools.

❑ Corn syrup or cream of tartar is often added to sugar solutions when making amorphous candy because each inhibits crystal formation. As mentioned earlier, corn syrup is composed of dissimilar sugar molecules which interfere with crystal formation. Cream of tartar, an acid, inhibits crystal formation by breaking sucrose into its constituent glucose and fructose molecules.

❑ When making crystalline candy, such as fudge and fondant, the goal is to create very small crystals. These tiny crystals give the candy a smooth, silky texture. To make fudge, a supersaturated sugar solution is allowed to cool slowly without being disturbed. When it nears room temperature and is extremely supersaturated, the solution must be stirred constantly until it is completely cool. Stirring encourages crystal formation and breaks any large crystals which have formed into smaller pieces.

❏ Most crystals form around a seed. Seeds may be either other sugar molecules or dust particles in the solution. The more supersaturated the solution, the higher the density of sugar seeds. The more seeds there are, the more crystals form simultaneously. The resulting crystals are relatively small in comparison to crystals formed in a less saturated sugar solution.

❏ If dust falls into the mixture, if the pot is bumped, or if there are undissolved sugar crystals on the sides of the pot, the mixture will crystallize around these unwanted seeds before the mixture is cool. The candy will have a grainy texture. To remedy this situation, the mixture must be stirred immediately to break up the large crystals. If the candy still has a grainy texture, the mixture can be reheated until it is again liquid and the cooling process can be carefully repeated.

❏ Candy-making uses a unique combination of art and science. An understanding of the properties of supersaturated sugar solutions and the factors that affect sugar crystallization can help to produce successful and delicious results.

Zebra Fudge

1¹/₂ cups sugar
³/₄ cup whipping cream
³/₄ cup half and half
1 tablespoon light corn syrup
5 ounces good quality milk chocolate, finely chopped
1¹/₂ ounces unsweetened chocolate, finely chopped
²/₃ cup chopped walnuts
2 ounces good quality white chocolate, finely chopped

Butter a 3- or 4-cup bowl
Line a small square pan with foil; a loaf pan will do
Preparation time: 2 hours
Yield: 12 1x2 inch pieces

1. In a medium-size saucepan, mix the sugar, cream, half and half, and corn syrup. Cook the mixture over medium heat until the sugar is completely dissolved. 2. Add the milk chocolate and unsweetened chocolate and stir slowly for about 20 minutes until the temperature is 235°F (113°C), or until a small amount forms a soft ball when placed in ice water. *As the mixture boils, the sugar concentration increases, and the boiling temperature rises. When the boiling temperature reaches 235°F (113°C), the sugar concentration is about 80 percent, which is the proper sugar concentration for making fudge.* 3. Pour the hot mixture into the buttered bowl and set it inside another larger bowl filled with ice water, taking care that no water gets into the mixture. *As the mixture cools, it becomes supersaturated with sugar and even the slightest movement or dust particle can cause the mixture to crystallize.* Let the mixture cool for about 20 minutes, until it is lukewarm (110°F, 43°C). 4. Beat the mixture with a wooden spoon for 4 to 5 minutes until it is almost cool. *Beating the mixture starts the crystallization process and breaks the crystals into small pieces, making the fudge smooth and creamy.* Add the walnuts and mix well. 5. Spread the mixture in the foil-lined pan and smooth out the top. 6. Melt the white chocolate in a small bowl in a warm oven or over a bowl of hot water. With a fork, drizzle the melted white chocolate on top of the fudge. 7. Refrigerate the fudge one hour before cutting it into squares. If the fudge does not harden sufficiently, put it back into a pot and reheat it until it reaches 235°F (113°C) or forms a soft ball when placed in ice water; then recool.

Variations:
1. Substitute ²/₃ cup peanut butter for the walnuts to make peanut butter ripple fudge. 2. Substitute ²/₃ cup white chocolate chips for the walnuts.

7 Starch

Starch ⋀⋀ is a molecule composed of a long chain of glucose molecules. It is produced and stored by many fruits, vegetables, grains, and nuts. These plants convert excess glucose into starch and store it in their roots or seeds. The starch stored in roots is converted into glucose again when the plant needs energy. Starch stored in the seeds is used to supply energy for growing seedlings. When we eat foods containing starch, our bodies break down the starch into glucose. Glucose is a source of energy for humans.

Two types of starch are found in plants. The first, called amylose, is composed of a single long chain of up to 600 glucose molecules. The second type, called amylopectin, is a branched molecule that contains up to 1,500 glucose molecules. Large collections of starch molecules, including both amylose and amylopectin, are packaged together into individual starch granules. The proportions of amylose and amylopectin in a starch granule depend on the specific plant from which the starch is taken. Within each granule, the separate starch molecules are held together loosely by weak bonds ⋀⋀⋀. The illustration on this page represents a single starch granule filled with many starch molecules.

The most common starches used in cooking are wheat flour, corn starch, potato starch, rice starch, arrowroot, and tapioca. When any starch is combined with cold water, the starch granules absorb water molecules ⌒⌒ and swell slightly. Heating the starch and water causes more water to be absorbed by the starch granule. The energy from the heat breaks the weak bonds that hold the starch molecules together. This allows water to enter the space between the individual starch molecules.

A starch granule expands as more water enters it. Each specific type of starch absorbs a characteristic amount of water. Some starch granules can absorb up to twenty-five times their own weight in water. When a starch granule has absorbed all the water it can, it is said to be gelatinized. In gelatinized starch, the original orderly arrangement of the starch molecules is disrupted by the water that enters the granule. This original pattern can never be restored. In addition, as the starch granules absorb more water, they begin to leak individual starch molecules into the surrounding solution. The drawing on this page shows this process.

❑ Starch is used to thicken water-based foods such as sauces, soups, pie fillings, and puddings. There are several reasons for its thickening ability. First, because starch absorbs water, it acts like a collection of small sponges in the solution. Because there is less free water in the solution, the solution is less fluid. Second, the large, swollen starch granules tend to slow the movement of the solution by interfering with the other components. Finally, the starch molecules that leak into the liquid become tangled with the large starch granules and act as baffles. These baffles immobilize the solution and make it thicker. Amylose starch immobilizes the solution more effectively because it is a long unbranched chain and acts as a bigger baffle. The illustration at right shows the structure of a starch-thickened sauce.
❑ When a solution of water and starch is cooled, the starch molecules that have been released from the granules begin to form random bonds with other starch molecules. This bonding results in the formation of a large network that holds the other ingredients in a semisolid gel. This is why a starch-thickened custard becomes thicker as it cools.
❑ Different starches have different properties. Flour, for example, is 70 percent starch by weight and has less thickening ability than does corn starch, which is 100 percent starch. In fact, only half as much corn

starch, compared with flour, is needed to produce the same amount of thickening. Different starches also create different gel consistencies. Corn starch creates a firm, smooth gel; flour creates a thick, pastelike gel.
❑ Several factors affect the thickening ability of starch. First, overheating or overstirring a starch solution breaks open the large starch granules that have absorbed water. This releases water back into the solution, making it thinner. Second, if the long starch molecules released into the liquid are

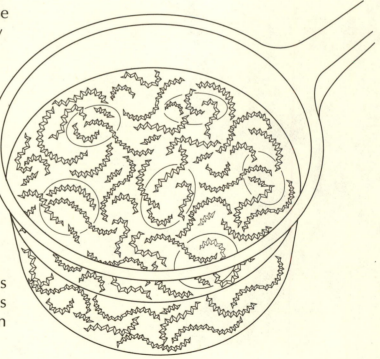

broken into shorter chains, their thickening ability is decreased. This can happen if an acid, such as lemon juice or cream of tartar, is added. It can also happen if the starch is cooked before it is used as a thickening agent. For instance, if flour is browned before it is added to a recipe, more flour must be used to obtain the same amount of thickening. Finally, sugar inhibits the thickening action of starch. In a solution containing a large quantity of sugar, most of the water molecules surround the sugar molecules, because sugar readily attracts water (is hygroscopic). Thus, less water is available to enter the starch granules. More cooking is needed to enable the starch granules to absorb enough water to thicken a sauce. (See Chapter 9, "Preserves," for a discussion of how the hygroscopic properties of sugar are used in the preparation of preserves.)

❑ Lumps will form in a sauce if the starch is not fully dispersed in a liquid before it is heated. A common example is lumpy gravy. When starch is added directly to hot water, the outer layer of a cluster of starch granules absorbs the hot water rapidly, forming a barrier that prevents water from reaching the center of the cluster. Starch lumps not only look and taste unpleasant but also are of little help in thickening the sauce. The illustration at left represents a lump of starch.

❑ There are several ways to prevent starch lumps, all of which involve dispersing the starch before it is added to hot liquid. First, the starch may be mixed with a small amount of cold water to ensure dispersion. Starch absorbs cold water less readily than hot water, so the granules are more easily separated. Second, the starch may be mixed with other dry ingredients before those ingredients are added to a hot liquid. Finally, the starch may be blended into fat, such as butter or oil because fat contains very little water. The starch can be dispersed without absorbing much water before being mixed into a hot solution.

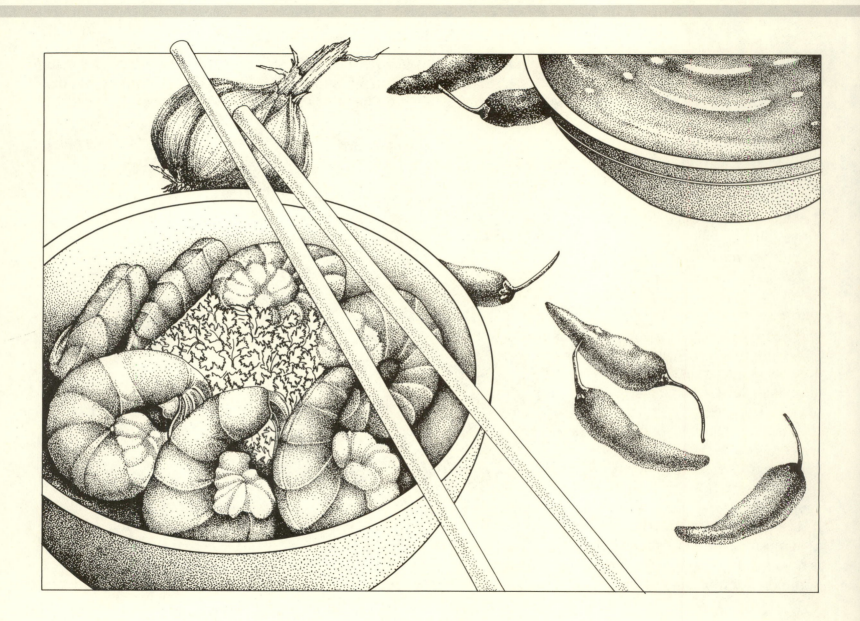

54

Thai Barbecue Sauce

2 tablespoons cornstarch
2 tablespoons cold water
1 cup sugar
1 teaspoon salt
¼ cup white wine vinegar
½ cup water
1 tablespoon catsup
2 to 4 serrano chilies (small, hot, green chilies)
2 cloves garlic, peeled and crushed

Preparation time: 30 minutes
Yield: about 1 cup

1. Blend the cornstarch and the cold water in a medium-size saucepan until the cornstarch is dissolved. *The starch must be dispersed in the cold water before it is heated to prevent the formation of lumps.*   **2**. Add the sugar, salt, vinegar, water, catsup, chilies, and garlic and mix thoroughly. Cook the mixture over medium heat until it boils.   **3**. Reduce the heat and simmer the sauce for about 30 minutes until it is thick and the flavors have blended. *The heat causes the starch granules to expand like sponges as they absorb water. They then release starch molecules into the liquid. The swollen granules and the free starch molecules partially immobilize the liquid and make it thicker.*   **4**. Serve immediately with barbecued chicken, shrimp, or egg rolls, or chill for later use.

8 Popcorn

❑ Few foods change as dramatically when cooked as does popcorn. In a split second, the small, hard pellets are transformed into fluffy, white puffs. What is popcorn and why does it pop?

❑ There are many varieties of corn that differ in the amount and type of starch and protein they contain. Sweet corn is eaten by people, and other types of corn are used as animal feed and for corn flour. An entirely different type of corn is used for popcorn. There are several varieties of popcorn: butterfly or snowflake popcorn explodes in all directions. Mushroom popcorn explodes in only one direction.

❑ A popcorn kernel is composed of starch granules ⟁⟁⟁, proteins, and a small amount of water ⟒⟓. The illustration below shows the structure of the starch granules within a popcorn kernel.

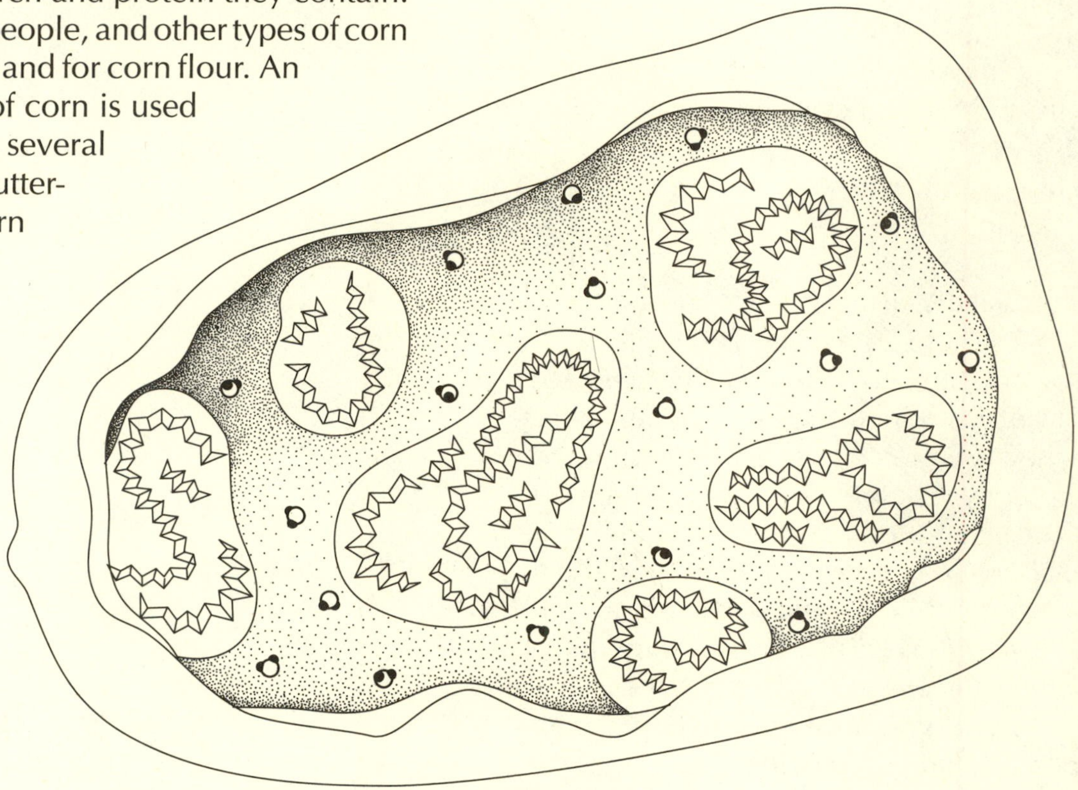

56

❏ Popcorn normally contains about 16 percent water. For corn to pop, the water content must be reduced to between 13.5 and 14 percent. Commercial popcorn producers dry the popcorn and seal it in airtight containers to preserve the correct moisture level.

❏ Heating popcorn causes the water molecules inside the kernel to move more quickly and to force their way between the starch granules. The drawing on this page shows water entering the starch granules. The starch granules then begin to gel. (See Chapter 7, "Starch," for a more detailed description of how this happens.) As the kernel is heated further, the water becomes a gas. Because water vapor takes up much more volume than does liquid water, the pressure inside the kernel increases.

58

❏ If the casing of popcorn kernels were porous, the pressurized water vapor would escape through the pores and the corn would not pop. But, the kernel casing is airtight and quite strong. As a result, the water vapor pressure builds up inside the kernel.

❏ A fundamental law of physics states that the pressure of any gas, at a given volume, increases as the temperature increases. Because the volume of a popcorn kernel remains constant while the temperature of the enclosed water vapor increases, the pressure inside increases. When the temperature inside the kernel reaches between 350°F (178°C) and 450°F (234°C), the pressure is so great that the kernel's casing bursts open. When the casing breaks, the contents of the kernel are no longer confined to a small volume. The starch and water vapor expand rapidly, and the expanded starch dries quickly, forming light, fluffy popcorn.

❏ If freshly made popcorn is left in a covered pot, the starch quickly reabsorbs the moisture and becomes soggy. That's why it is important to quickly transfer freshly made popcorn to an open bowl, where all the moisture can evaporate.

❏ By using the simple relationship between temperature, pressure, and volume, it is possible to predict how gases will react under many different conditions. Popcorn is just one example of

how the fundamental gas laws are used in the kitchen. These laws are also applied when leavening batters and when using a pressure cooker.

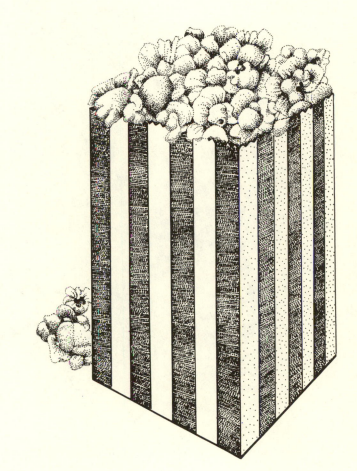

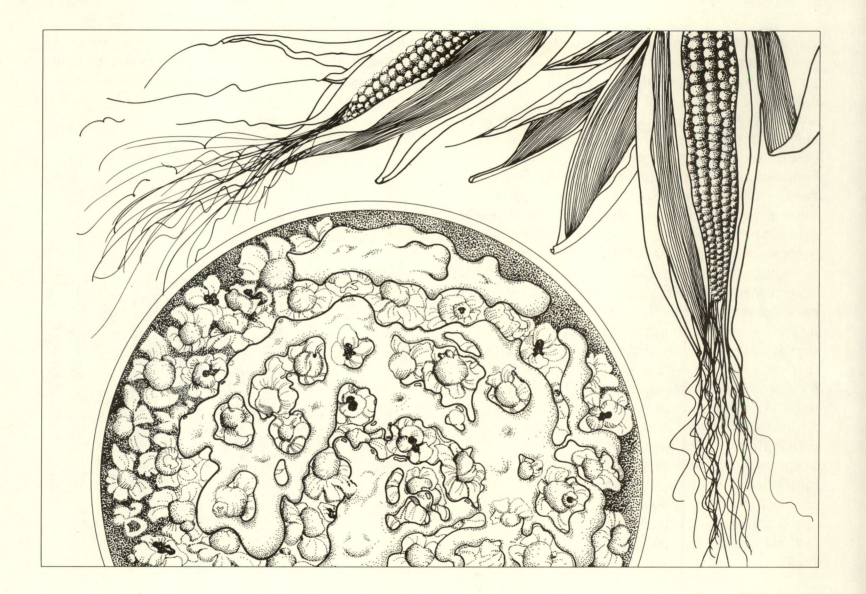

Popcorn Nachos

2 tablespoons vegetable oil
1/4 cup popcorn kernels
1 cup grated cheddar cheese
2 tablespoons diced mild green chilies
1 teaspoon chili powder

Preheat oven to broil
Preparation time: 20 minutes
Yield: 4 servings

1. Heat the oil in a medium-size covered pot with 2 or 3 popcorn kernels. When the kernels pop, add the remaining kernels and cover the pot.   **2**. Shake the pot over the flame until the popping stops. *Popcorn pops when the pressure from the hot water vapor inside the kernels becomes so great that it bursts the corn casing.*   **3**. Quickly pour the popcorn into a shallow ovenproof serving dish. *If the popcorn is left in a covered pot after it has popped, the starch will reabsorb the moisture, and the popcorn will become soggy.*   **4**. Sprinkle the popcorn with grated cheese, then the chilies, and finally the chili powder. Broil for a few minutes until the cheese melts.   **5**. Serve immediately.

Variations:
1. Add grated Monterey Jack cheese as well as cheddar cheese.   **2**. Sprinkle the popcorn with diced avocado after taking the popcorn out of the broiler.

9 Preserves

❏ Preserves, including jams and jellies, have been made for hundreds of years to prepare seasonal fruits so that they can be enjoyed year-round. Although preserves are usually made from a few simple ingredients (fruit, water, and sugar), the chemistry involved is relatively complex. This complexity is due to the reactions involving pectin, a molecule extracted from the fruit that causes preserves to gel.

❏ Pectin /\/\/\/\ is a long molecule composed of a chain of smaller molecules. Each of these smaller molecules has a structure similar to that of sugar. Many types of fruit contain pectin, which functions like cement within the walls of the cells. When making preserves, the fruit is boiled to extract the pectin from the cells. Once the pectin is free of the cells, the individual long pectin molecules can bond to one another, forming a continuous network that

traps the other ingredients in the solution. The result is a gel similar to that formed by starch or gelatin.

❏ Because pectin molecules are negatively charged, they normally repel one another. Individual pectin molecules are also surrounded by water molecules ❑ ❑, which are attracted to the negative charge of the pectin. Because the pectin molecules repel one another and are surrounded by water, they rarely interact with one another under normal conditions. The drawing on this page shows a pectin molecule surrounded by water molecules.

❑ For pectin molecules to interact with one another, the negative charge on the molecules must be neutralized. This is done by making the solution acidic (pH of 2.8 to 3.4), by using acidic fruit, or by adding lemon juice. When making preserves with grapes or tart apples, which are naturally acidic, there is no need to add anything else. If you are using fruit that is not acidic, such as sweet apples or bananas, it is necessary to add acid in the form of lemon juice.

❑ Once the charge on the pectin molecules is neutralized, the water molecules surrounding the pectin must be moved aside to allow the pectin molecules to come into close contact. This is done by raising the sugar concentration to 60–65 percent. Because sugar

attracts water (is hygroscopic), the water is drawn away from the neutralized pectin molecules and gathers around the sugar molecules instead.

❑ These steps allow the pectin molecules to associate with one another to form a network enclosing the other ingredients. If there is not enough acid or sugar in the solution, the pectin will not gel, and the result will be syrup instead of jam.

❏ The pectin concentration within preserves affects the consistency of the final gel. Too little pectin causes runny preserves; too much pectin causes a rubbery consistency. Some fruits, such as grapes, berries, green apples, and citrus fruits, naturally contain enough pectin to form a proper gel. Other fruits, such as peaches, apricots, and strawberries, require the addition of prepared pectin to achieve the proper consistency. Prepared pectin is extracted from fruits having high pectin concentrations, such as apple skins and cores or the white, spongy layer under citrus fruit skin.

❏ Ripe fruit contains the highest concentrations of pectin. Unripe fruit contains protopectin which has no gelling ability. As fruit ripens, the protopectin is converted into pectin by enzymes in the fruit. The conversion of protopectin to pectin is responsible for the softening of fruit as it ripens. When fruit becomes overripe, the pectin turns into pectic acid, which also cannot gel. Therefore, it is best to use barely ripe fruit when making preserves; this maximizes the pectin concentration in the solution.

❏ Making preserves is simple if the pectin, sugar and acid concentrations are correct. First, the fruit is briefly simmered to extract the pectin from the cells. Then an acid (lemon juice) is added to the solution, if needed, to neutralize the pectin molecules. Finally, sugar is added to attract the water molecules away from the pectin molecules. The solution is then boiled until the sugar is dissolved and reaches a concentration of 60 to 65 percent. A candy thermometer can be used to determine the sugar concentration. (See Chapter 6, "Sugar," for a description of how to measure the sugar concentration with a candy thermometer.)

❏ It's very important not to overcook preserves. Overcooking leads to many problems, including caramelization of the sugar, excessive evaporation of water, loss of the bright color of the fruit, and evaporation of the flavor molecules. Prolonged cooking can also lead to unwanted sugar crystals and can cause the pectin molecules to break down. When preserves have been cooked just long enough to dissolve the sugar, they should be spooned or poured quickly into clean jars.

❏ Jams and jellies preserve the fruit because any bacteria that enter the highly concentrated sugar solution die quickly. Bacteria are normally filled with a salt and sugar solution that is much less concentrated than that of the preserves. When bacteria enter the preserves, the water inside the bacteria flows out across the bacterial cell membrane into the surrounding solution by a process called osmosis. Osmosis is defined as the tendency for water to flow from an area of lower concentration of dissolved particles to an area of higher concentration of dissolved particles. As a result of osmosis, bacteria dry out and die if they enter a highly concentrated sugar solution.

❏ Molds, however, can grow on the surface of preserves. If water evaporates from the surface of the
preserves and then condenses, this pooled water has
a much lower concentration of sugar than does the
preserves. The condensed water supports
the growth of molds. Covering the
surface of preserves with paraffin
wax prevents evaporation of
the water and thus pre-
vents mold growth.

66

Tangerine Jam

8 tangerines
2 lemons
Water
Sugar (about 4 to 6 cups)

Preparation time: 1½ hours
Yield: About 1 cup

1. Massage the fruit for a few seconds to help release the juice. **2.** Cut the fruit in halves; squeeze the juice into a measuring cup and set aside. With a spoon, scrape out the membranes and the pulp and discard them. **3.** Put the rinds into a heavy-bottomed stockpot and cover them with cold water. *The rinds are the source of pectin, which makes the jam gel.* **4.** Simmer the rinds until they can be pierced easily with a fork, about 20 to 30 minutes. The tangerine rinds will soften sooner than the lemon rinds. Remove the softened rinds and discard the water. *The boiling water disrupts some of the cells within the rind and allows the later release of pectin molecules.* **5.** Slice the softened rinds into ½ inch strips and add them to the juice in the measuring cup. Note the volume. **6.** Pour the juice and rinds into the pot and add an equal volume of sugar. Cook the mixture over medium heat while stirring until the jam starts to thicken, about 30 minutes. *The pectin comes out of the rinds and enters the sugar solution. In an acidic solution that contains a high sugar concentration, the pectin forms a gel.* **7.** When the mixture is so thick that a small droplet forms a soft ball when placed in cold water, the jam is done. Remove the rinds with a slotted spoon and discard them. **8.** Pour the hot jam into a clean jar and seal it. **9.** Serve on bread or use as a glaze for chicken or turkey.

Variations:
1. Use any combination of citrus fruits. **2.** Grate the rinds instead of slicing them and do not remove them from the jam — this makes marmalade.

Solutions

10 Water

The boiling point of water ◯◡ is the temperature at which it bubbles and turns into a gas. But what determines this temperature? The answer is found by examining the various forces that act upon any liquid.

The molecules within a liquid are always moving. Some of them move fast enough to escape from the surface of the liquid and become a gas. This process is called evaporation. The warmer the temperature of the liquid, the more molecules escape from its surface and the faster the liquid evaporates.

The pressure exerted on the surface of a liquid by the escaping molecules is called the vapor pressure ⬆ . There is also a constant downward pressure on the surface of a liquid from the air in the atmosphere, called the atmospheric pressure ⬇ . This is the total weight of the column of air over the liquid, beginning at the surface of the liquid and extending all the way through the earth's atmosphere. At room temperature, the atmospheric pressure is greater than the vapor pressure of a liquid and, along with gravity, holds the liquid in the container. The drawing on this page illustrates a pitcher of water. The two up-ward arrows indicate the vapor pressure, and the three downward arrows indicate the atmospheric pressure.

❏ Heat causes the molecules within a liquid to move faster which, in turn, causes more molecules to escape from the surface of the liquid. As a result, the vapor pressure increases. When the vapor pressure is equal to the atmospheric pressure, the liquid boils. The drawing on this page illustrates this process.

❏ Every liquid has a specific boiling temperature. Once a pure liquid reaches this temperature, it cannot get any hotter, no matter how much heat is applied. Cooking food in a boiling liquid over a higher flame will not cook the food faster, though the water will boil away more quickly.

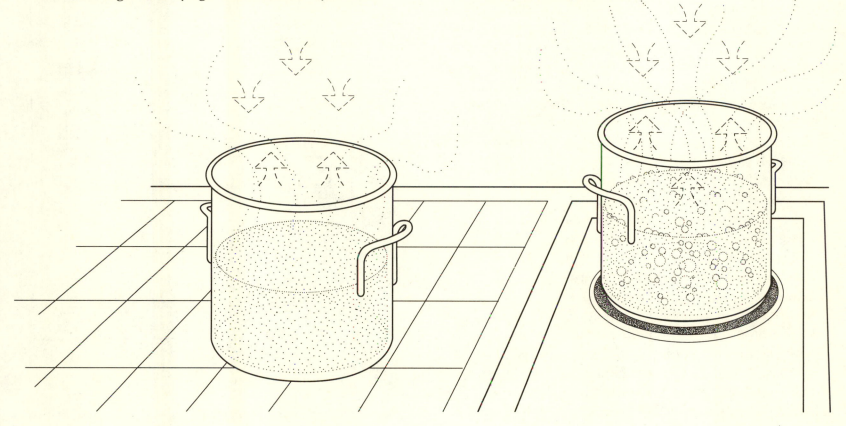

❑ The boiling temperature of a liquid can be increased or decreased by changing the atmospheric pressure or the vapor pressure. Water, for example, boils at 212° F (100° C) at sea level. At higher altitudes the atmospheric pressure is lower. Therefore the vapor pressure of water equals the atmospheric pressure (and the water boils) at a lower temperature. The boiling temperature of water decreases 1°F for every 500-foot increase in elevation above sea level. The drawing below shows that the atmospheric pressure on a mountaintop is lower than that at sea level.

❑ People living at high altitudes often use pressure cookers to compensate for the lower boiling temperature of water. A pressure cooker is an airtight container that traps the steam produced during heating. The pressure inside the cooker increases as the water inside evaporates. A pressure cooker, in effect, artifically raises the atmospheric pressure inside the pot. The water in the pot must be heated to a higher temperature for the vapor pressure to equal the higher atmospheric pressure. Food immersed in the boiling water within a pressure cooker cooks more quickly. Pressure cookers can also be used at sea level to boil food at a temperature above 212°F (100°C).

❏ Dissolved particles, such as salt or sugar, also increase the boiling temperature of water. For example, when salt dissolves in water, it separates into positively charged sodium ions •• and negatively charged chloride ions ○. These ions cannot evaporate. Individual water molecules each have a negatively and a positively charged side. The negatively charged side is attracted to the sodium ions, and the positively charged side is attracted to the chloride ions. The water molecules surround the ions, as seen in the illustration at right.

❏ Water molecules that surround the sodium and chloride ions are not free to escape from the surface of the liquid, so the vapor pressure of the liquid decreases. The solution then must be heated to a higher temperature to increase the vapor pressure so that it matches the atmospheric pressure. The more dissolved particles there are in the water, the higher the boiling temperature.

❏ A simple pot of water illustrates how vapor pressure and atmospheric pressure affect the boiling temperature of a liquid. These forces, which must be equal for a liquid to boil, can be modified by bringing the pot of water to the top of a mountain, by using a pressure cooker, or by dissolving salt or sugar in the water.

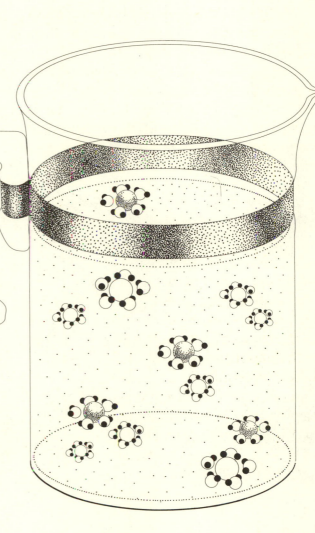

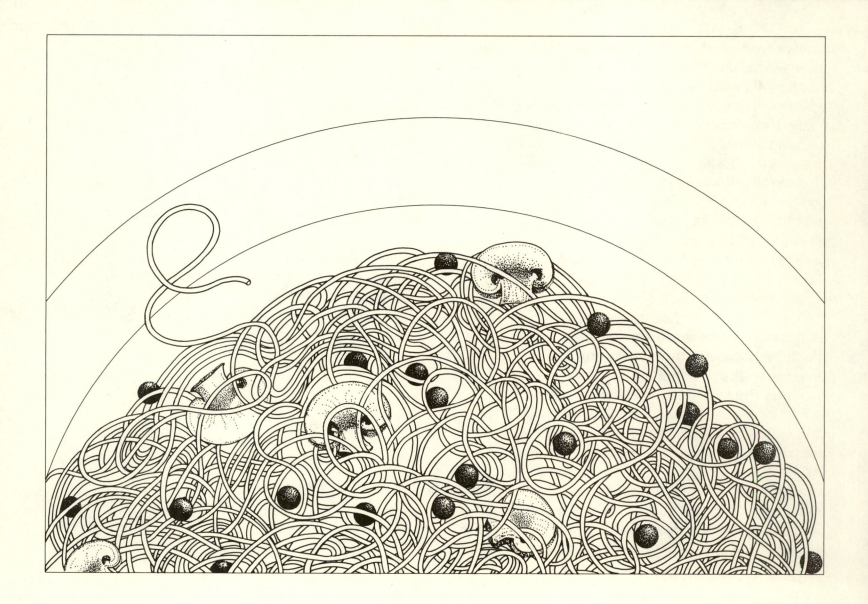

74

Prima Pasta

1 teaspoon salt

1/4 cup olive oil

4 cloves of garlic, peeled, crushed, and chopped

1 large red pepper, stem and seeds removed, sliced into long, narrow strips

1 large zucchini, sliced into long, narrow strips

1/4 pound prosciutto (or ham), sliced paper thin and shredded into small pieces

1 cup fresh or frozen peas

Freshly ground pepper

1 teaspoon dry basil or 2 tablespoons chopped fresh basil

1 pound fresh fettucine (pasta)

1/2 cup grated Parmesan cheese

Preparation time: About 20 minutes

Yield: 2 servings as a main course, 4 servings as a first course

1. Add the salt to a large pot of water and bring the water to a boil. *The salt will increase the boiling temperature of the water, but not to any significant degree.* **2**. Heat the olive oil in a skillet and add the garlic. Cook until soft, about 4 minutes. **3**. Add the red pepper to the oil and garlic and cook until soft, about 5 minutes. Add the zucchini and cook 4 or 5 minutes more. **4**. Add the prosciutto and sauté about 4 minutes. Add the peas and sauté until they are heated through, about 2 minutes. **5**. Stir in the pepper and basil and set aside. **6**. Add the fettucine to the boiling water and cook until tender. Drain the fettucine and put it on individual plates. **7**. Top the pasta with a large portion of the sautéed vegetables and prosciutto. **8**. Sprinkle with grated Parmesan cheese and serve immediately with a tossed salad and fresh bread.

Variations:

1. Use mushrooms, or sliced black olives instead of prosciutto. **2**. Use freshly steamed, sliced asparagus instead of peas. **3**. Add 1/2 cup chopped onion at the same time as the garlic.

11 Coffee

❏ Coffee is one of the most popular beverages in the world. Some people drink strong espresso; others prefer their coffee diluted with milk and sweetened with sugar. Many people drink coffee for the stimulating effects caused by its caffeine content.

❏ Coffee begins as the pit of the coffee tree fruit, which grows in many areas of the world, including South America, Kenya, Hawaii, and the East and West Indies. The cherrylike fruit of the coffee tree is harvested when it has ripened to a red or deep purple color. Each fruit yields from one to three pits that ultimately become coffee beans. The drawing on this page shows the fruit of the coffee tree before it is picked.

❏ As the fruit ages, it is softened by the action of its own enzymes. After a pulping machine separates the fruit from the pits, the pits are dried. The dried pits, or beans, are then shipped to countries around the world, where they are roasted.

❏ Unroasted green coffee beans contain complex sugars, starch, oil, protein, fiber, caffeine, water, and tannins. Tannins are plant pigments that make coffee, as well as tea, wine, and unripe fruit, astringent. (This property of tannins is discussed further in Chapter 12, "Tea.")

❏ Roasting causes many dramatic changes in the green coffee beans. First, about 16 percent of the moisture remaining in the beans is removed, making the beans brittle

and porous. The high roasting temperature also promotes reactions between the sugars and the proteins within the beans, turning the beans brown. Finally, roasting breaks down many of the molecules within the beans and causes other molecules to evaporate. The longer coffee beans are roasted, the drier and darker they become.

❏ After the beans are roasted, they are ground. Hot water is then poured over the beans to extract the compounds that give coffee its flavor, aroma, and color. The finer the beans are ground, the more surface area there is from which the water can absorb the compounds. For example, strong espresso is made by forcing steamed water through a very fine grind of coffee. A coarser grind is used for filtered coffee, which has a much milder flavor. The drawing on this page shows an old-fashioned coffee grinder.

❏ The complex flavor and aroma of coffee result from more than 300 different molecules. Some of these molecules are volatile and evaporate when the beans are exposed to air. Other molecules react with oxygen in the air and change the flavor of the coffee. Evaporation and oxidation occur more rapidly in ground coffee than in whole beans because ground beans have a larger surface area. Thus it is better to store whole coffee beans rather than ground beans. If ground beans must be stored, it is better to keep them frozen until they are used, both to prevent the volatile molecules from escaping and to inhibit oxidation. Coffee made from stale beans tastes bitter and flat.

❏ There is a science to making a perfect cup of coffee. When hot water contacts the ground beans, the acids are drawn out of the beans first. Weak coffee, which is made by passing the water very quickly over a small amount of ground coffee, tastes sour. Overbrewing, on the other hand, overexposes the coffee to the hot water and releases many bitter tasting molecules. For these reasons, ground coffee should be in contact with hot water for the correct amount of time. This is frequently accomplished by using a coffee filter and funnel and the proper amount of ground beans and water. A typical filtered coffee maker is shown at right.

❏ As a rule of thumb, it is best to use about one tablespoon of ground coffee for each cup of water, and to use freshly boiled water to retain the dissolved oxygen. Once the water boils, let it cool to about 200°F (94°C) before pouring it over the coffee. Once coffee is prepared it should not be reheated because boiling or reheating coffee drives off many of the aromatic molecules, leaving a flat-tasting brew.

❏ The caffeine in coffee is a stimulant and has many effects on the body. In small doses, caffeine can initially increase the ability to concentrate, increase alertness, decrease reaction time, and increase muscle strength. After these desireable effects wear off, however, a person may feel tired and nervous.

❏ Caffeine also stimulates digestive system secretions, increases the heart rate and blood pressure, increases the metabolic rate, enlarges the blood vessels in most of the body, and constricts the blood vessels in the brain. In addition, consuming caffeine can cause problems such as insomnia, heartbeat irregularities, muscle tremors, nervousness, stomach irritation, fatigue, and headaches. In short, caffeine has complex effects on many systems of the body.

❏ Because of the undesirable effects of caffeine, many people drink decaffeinated coffee. Coffee beans are decaffeinated by two primary methods. Some manufacturers steam the unroasted beans and then add a solvent to extract the caffeine. Water is used to wash the solvent and the caffeine out of the beans. The beans are then dried and roasted. Other manufacturers use the water-decaffeination method. The beans are steamed and the caffeine-rich exterior is scraped off. Both processes remove some of the flavorful compounds along with the caffeine.

❏ Each step in the process of preparing coffee is important, from roasting the beans to brewing the coffee. The goal is to create a perfect blend of flavors, aroma, warmth, and stimulation.

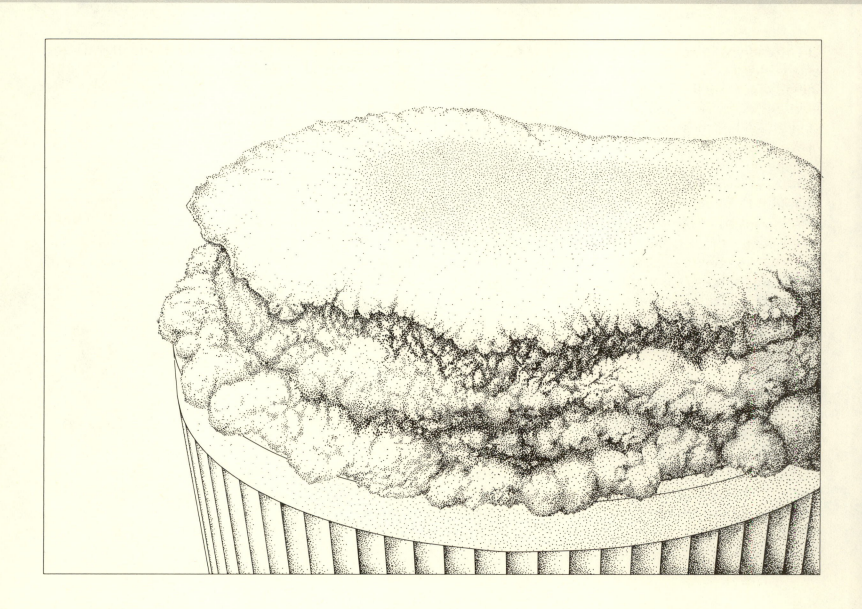

Coffee Soufflé

2 tablespoons ground coffee, preferably espresso
1 cup water
3 tablespoons butter
3 tablespoons flour
1/4 cup heavy cream
5 eggs, separated
1/4 cup sugar
A pinch of salt
1 teaspoon vanilla extract
1/4 teaspoon cream of tartar

Butter a 6 cup soufflé dish and sprinkle it with sugar.
Preheat oven to 350°F
Preparation time: 1 1/4 hours
Yield: 4 servings

1. Put a coffee filter into a funnel and add the coffee.   **2.** Boil the water and let it cool for several seconds. Pour the water over the coffee and let it drip into a cup. Set the coffee aside. *The ground beans release hundreds of different molecules into the hot water.*   **3.** In a small pot, melt the butter and stir in the flour. Cook until smooth. *Dispersing the flour in the melted butter prevents the formation of lumps when hot ingredients are added.*   **4.** Add 3/4 cup of coffee slowly. Then add the cream and continue cooking the mixture over medium heat until it thickens or forms soft peaks when stirred, about 5 to 7 minutes.   **5.** Remove the mixture from the heat and set aside to cool.   **6.** Beat the egg yolks with the sugar and salt. Slowly pour the warm coffee mixture into the yolks while beating the mixture. Then add the vanilla extract.   **7.** In a separate bowl, whip the egg whites with the cream of tartar until they are stiff. *The whipped egg whites are stabilized by the acidic cream of tartar.* Fold the egg whites into the yolks and pour the mixture into the prepared soufflé dish.   **8.** Bake in a 350°F oven for 40 to 50 minutes, until the soufflé is slightly browned on top. Sift powdered sugar on the top, if desired. Serve immediately.

12 Tea

❏ From the British afternoon tea to the highly ritualized Japanese tea ceremony, tea has a place in the traditions of many societies. Despite the diversity of cultures surrounding tea, all tea is produced from the leaves of the tea plant, *Camellia sinensis.* It is a flowering evergreen shrub, as illustrated on this page. The tea plant thrives in tropical climates, especially at high altitudes, and requires an average of 100 inches of rain a year. The differences between the many types of tea arise from where the tea is grown, how and when it is picked, and how it is processed. (Herbal teas, which are composed of dried leaves such as mint or flowers such as chamomile, are not true teas.)

❏ Tea leaves are carefully harvested by hand. The best tea comes from the youngest leaves and unopened leaf buds, which contain the highest concentrations of enzymes, caffeine, and tannin.

❏ Tannins  are pigment molecules found in many types of plants. They provide a valuable defense mechanism for plants because they bind readily to proteins. In living plants, tannins bind to the surface proteins of bacteria that infect the plants and kill the bacteria. Because tannins also bind to the surface proteins of animal skins, they are also used to turn skins into leather. The tannins found in tea, coffee and wine bind to surface proteins in the mouth, causing the astringency of these beverages.

❏ Black tea is made by drying the leaves in the sun for up to 24 hours. When they become withered and pliable, they are rolled in large rolling machines that bruise and crush the leaves. Rolling exposes the cells to oxygen and gives the leaves the twist that is characterisic of high-quality tea.

❏ As soon as the tea leaves are rolled, they begin to "ferment." This process is not true fermentation because there are no microorganisms involved. Tea fermentation occurs when the tannins come into contact with enzymes in the leaves and oxygen in the

air. In the presence of oxygen, the enzymes transform the colorless, flavorless tannin molecules into flavorful dark tannins. The drawing above illustrates the twisting of the tea leaves and the darkening of the tannin molecules that begins when the leaves are rolled.

❏ There are several intermediate stages in the fermentation reaction, producing first yellow, then red, and finally, brown tannins. The yellow tannins have the most flavor; the darker tannins contribute primarily to the dark color of the tea.

❏ When fermentation has produced the proper balance of yellow, red, and brown tannins, the leaves are heated to denature the enzymes, which stops the reaction. The heat also causes some carmelization of the leaves, turning the leaves black. Once the leaves are dried they are separated into groups of different sizes, using sieves with progressively smaller holes. All leaves used to prepare a single pot of tea must be of similar size to produce tea with a consistent flavor.

❏ Different types of tea are made by varying the process. Green tea, for example, is made by inactivating the enzymes to prevent fermentation. The leaves are steamed, rather than withered, before the rolling and drying process begins. Because the pigment molecules are not developed, the leaves remain green and produce a light tea with a greenish-yellow hue.

❏ Oolong tea is produced by fermenting the leaves for only a brief time. This leads to a light brown tea with a greenish hue. The flavor of oolong tea is stronger than that of green tea and milder than that of black tea.

❏ Teas also differ by the size and twist of the processed leaves. Larger leaves with more twist tend to release their color and flavor more slowly, resulting in a mellower tea. Smaller, less twisted leaves produce a darker, stronger tea. Tea leaves are graded according to their size and the quality of the original leaves. From highest to lowest quality, leaves are rated as broken orange pekoe, orange pekoe, pekoe, pekoe souchong, souchong, fannings, and dust. The smaller pieces are usually used in tea bags.

❏ Tea leaves are also classified by where they are grown, including Ceylon (now Sri Lanka), Darjeeling (a district of India), Japan, China, Indonesia, and Taiwan. Various flavorings, such as clove, orange or lemon peel, cinnamon, peppermint and chamomile, as illustrated below, are often added to tea. Oil of bergamot, a type of orange, is used to flavor Earl Grey tea. Tea can be processed in slightly different ways and can be blended to form variations.

❏ When brewing tea, it is best to start with cold water. The water should be poured over the tea leaves just as it begins to boil. Using freshly boiled water ensures that the tea leaves unwrap completely and release their flavorful tannins into the water. The drawing on this page shows water being poured over tea leaves in a strainer. Once the water is in the cup, the leaves can be steeped as long as desired.

❏ Because the first molecules extracted from the tea leaves are the flavorless brown tannins, short steeping times result in brown, flat-tasting tea. Longer steeping times release the more flavorful yellow tannins and the caffeine. If the leaves are left in the water too long, however, too much caffeine is released, resulting in bitter tea. The optimum steeping time for a perfect blend of color and flavor is usually between 3 and 5 minutes. If you prefer weak tea, it is better to dilute normal strength tea with hot water rather than to steep the tea for a briefer time. If you prefer strong tea, it is better to use more tea leaves and a normal steeping time rather than to steep fewer leaves longer.

❏ The yellow tannins give tea its astringency because they bond to the surface proteins in your mouth. If milk is added to tea, the milk proteins bind to the tannins. As a result, the tea is much less astringent.

❏ Dry tea leaves contain as much as 4 percent caffeine, by weight, which is much more than coffee beans contain. However, prepared tea usually contains less caffeine than does coffee because much less tea is used to make the final beverage. (See Chapter 11, "Coffee," for a description of the effects of caffeine.)

Indian Spiced Tea

2 cups water
1 tablespoon chopped fresh ginger root
¾ teaspoon ground cinnamon
12 whole cloves
¼ teaspoon ground nutmeg
1 teaspoon black tea
2 teaspoons sugar
1 tablespoon cream
½ teaspoon lemon

Preparation time: 15 minutes
Yield: 2 servings

1. Put the water, ginger, cinnamon, cloves, and nutmeg into a medium-size pot. Cover the pot and boil the mixture for 10 minutes.   **2.** Turn off the heat and add the loose tea, sugar, cream, and lemon. Steep for 3 minutes. *As the tea leaves unfold, the first molecules released are the dark pigment molecules followed by the flavorful tannins.*   **3.** Strain the mixture into cups and serve immediately.

13 Ice Cream

❏ From both a scientific and an epicurean stand-point, ice cream is a fascinating food to study. The process of making ice cream uses several basic scientific principles, and the result is a delicious dessert. In this chapter, the process of making ice cream is used to present the basic scientific principles of freezing and melting.

❏ Cream, sugar, and selected flavorings are often the only ingredients in homemade ice cream. These ingredients are mixed together and then precooled to speed the freezing process. The cooled mixture is placed in the cannister of an ice cream freezer which, in turn, is placed inside a large tub. The tub is filled with a brine of ice and salt that cools the mixture. A paddle, or dasher, inside the cannister is turned manually or electrically to whip air into the ice cream and to break large ice crystals into smaller pieces. The drawing on this page shows a manual ice cream maker and the ingredients necessary to make ice cream.

Two major structural changes occur when liquid ice cream freezes. First, most of the water in the cream freezes into small ice crystals. Second, air bubbles are incorporated into the mixture. The goal is to create a smooth, creamy mixture of microscopic ice crystals ☆☆☆, air bubbles ▨, fat globules ⬭, and sugary liquid, as shown in the illustration at right.

In general, dissolved particles, such as salt or sugar, lower the temperature at which a liquid freezes. The effect of dissolved particles on the freezing and melting temperatures is called freezing point depression. The more dissolved particles there are, the greater the freezing point depression. Because ice cream contains a very high concentration of dissolved sugar, its freezing point is below that of water, 32°F (0°C). Freezing point depression also occurs in the brine, which has a high concentration of dissolved salt.

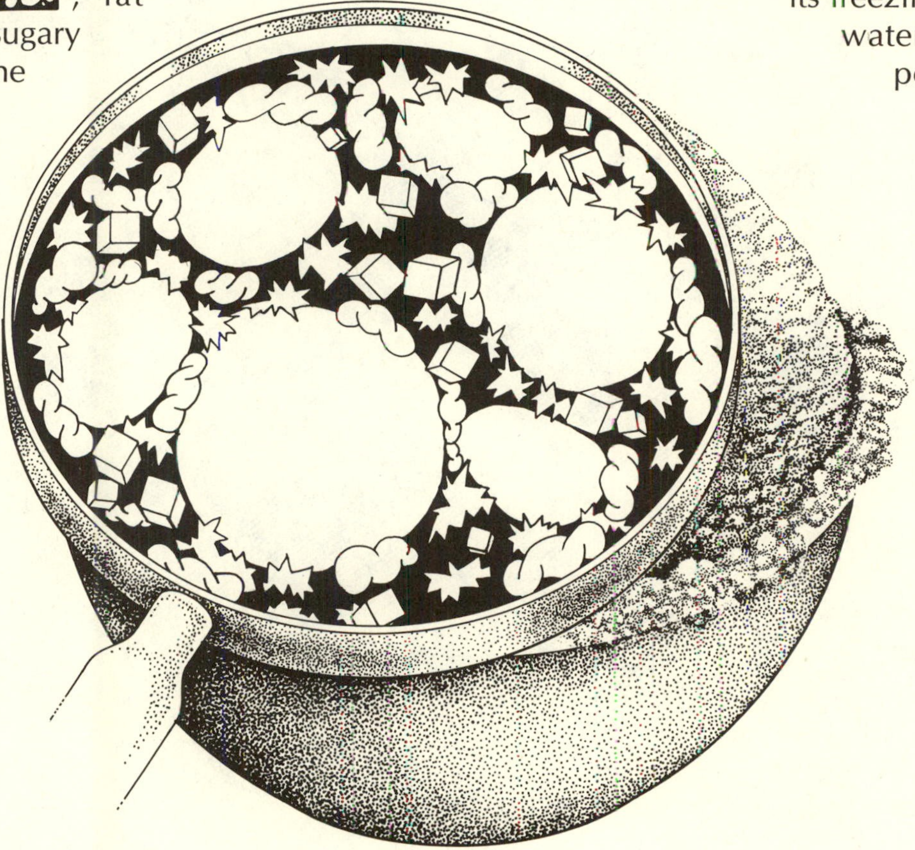

❏ Pure water ❷❛ freezes into ice crystals, which have a repetitive structure. (See Chapter 6, "Sugar," for a discussion of crystal formation.) To create ice crystals in a solution with a high concentration of dissolved sugars or salt, the temperature of the solution must be lowered to overcome the interference of the particles. The drawing on this page illustrates how pure water freezes into a crystalline structure. The drawing on the facing page illustrates how dissolved sugar interferes with ice crystal formation by getting in the way of the water molecules.

❏ Once the ice crystals form, water molecules on the surface of the ice constantly detach from and reattach to the surface of the crystals. Some of the ice melts, while some water refreezes. If the melting and refreezing occur at the same rate, the amount of ice remains constant.

❏ If salt or sugar are added to the surface of ice, they obstruct the water molecules that rejoin the crystals, inhibiting the refreezing process. The dissolved particles do not , however, interfere with the melting of the ice. As a result, there is more ice melting into water than there is water refreezing into ice. Eventually the ice turns to water. This is why ice, in the presence of salt or sugar, melts at temperatures at which it would normally remain frozen. As more salt is added to ice, the melting temperature drops further, leading to even colder water. (This same principle is used when roads are salted to melt the winter ice.)

❏ There is a direct relationship between the amount of sugar in ice cream and the amount of salt needed in the brine. The higher the concentration of sugar in the ice cream, the lower the freezing temperature of the water it contains. Therefore, more salt is needed in the brine to decrease the freezing temperature of the ice in the tub.

❏ Ice crystals form in the sugar solution as the ice cream cools. Since the formation of ice crystals removes water from the solution, the remaining unfrozen solution has a higher sugar concentration and, therefore, a lower freezing temperature. Eventually, the remaining liquid becomes so concentrated with sugar that it never freezes.

❏ As the ice cream cools, it is mixed with a paddle to break the ice crystals into small pieces and to incorporate air. The drawing on this page shows the inital formation of large ice crystals in the ice cream. The drawing on the facing page illustrates how the motion of the paddle breaks the crystals into smaller pieces.

❏ The motion of the paddle also whips air into the ice cream mixture, making it much lighter. This added air is called overrun. If the volume of ice cream doubles as a result of added air, the ice cream is said to have an overrun of 100 percent. Homemade ice cream usually has an overrun of 30 to 50 percent, because it is difficult to whip air into ice cream using a home ice cream maker. Commercial ice cream often has a much higher overrun, because manufacturers can easily pump air into the ice cream as it freezes. In general, the lower the overrun, the higher the quality of the ice cream.

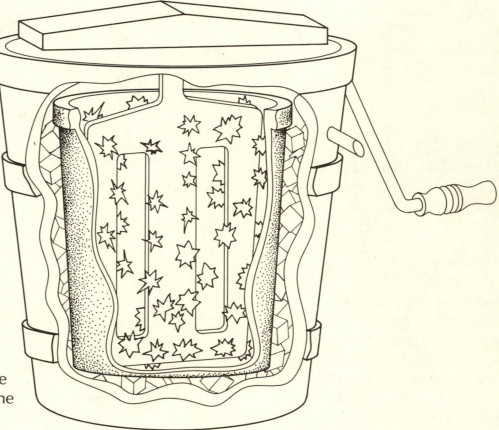

❑ Ice cream also contains milkfat which serves two purposes. First, by coating the small ice crystals, the fat inhibits the formation and growth of large crystals. Second, the fat makes the ice cream feel rich and creamy. Higher quality ice cream tends to have a higher milkfat content. Commercially-made ice cream is usually homogenized in order to distribute the fat evenly throughout the mixture, resulting in a consistent texture.

❑ The last variable in making ice cream is the flavor, which can be modified endlessly. Many flavors have been tried, from sweet vanilla to sweet potato, and from fudge marble to marbled garlic. However, since there are certainly many unique combinations that have not been tried, there is no limit to the flavors you can create.

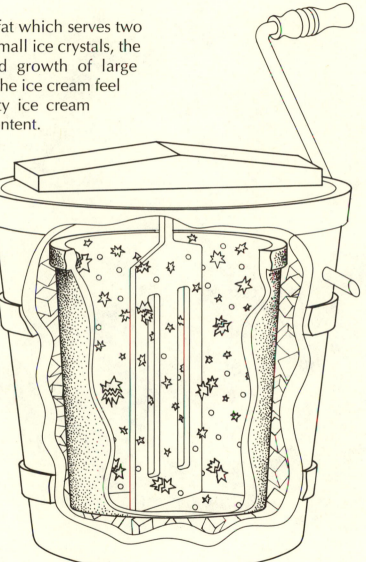

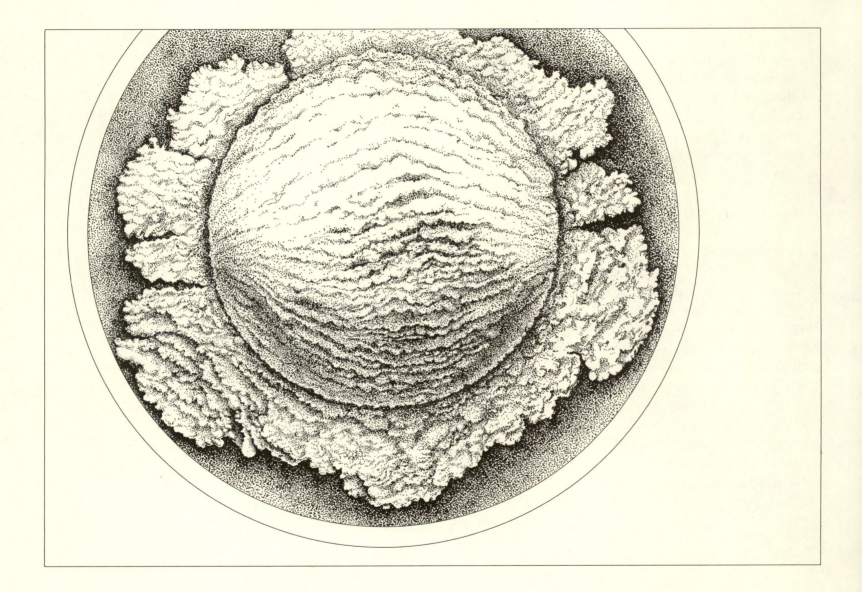

Pumpkin Ice Cream

2 cups milk
4 cups heavy cream
1/2 teaspoon ground cinnamon
1/4 teaspoon nutmeg
6 egg yolks
1 2/3 cups sugar
1 16-ounce can of pumpkin purée
crushed ice
rock salt
water, for the ice cream maker

Preparation time: 2 3/4 hours
Yield: 2 1/2 quarts

1. In a 4-quart saucepan, scald the milk, 1 cup of cream, the cinnamon, and the nutmeg. **2.** In a large bowl, beat the egg yolks together with 1 cup of sugar. Gently stir the hot milk into the egg yolks and then pour the mixture back into the saucepan. **3.** Cook this custard over low heat, stirring it constantly with a wooden spoon. *The heat denatures the egg proteins; they then coagulate and thicken the custard.* **4.** When the custard is thick enough to coat the back of the wooden spoon, about 10 minutes, strain it into a large clean bowl and add the remaining cream. Refrigerate the custard for at least 1 hour. *If the mixture is cool, it freezes more quickly in the ice cream freezer.* **5.** In a separate bowl, blend together 2/3 cup sugar and the pumpkin puree. **6.** Add the sugar and pumpkin mixture to the chilled custard and mix well. Put the mixture in the metal cannister of an ice cream maker. **7.** Fill the tub surrounding the cannister with layers of crushed ice, rock salt, and a small amount of water. *The more sugar there is in the ice cream mixture, the lower its freezing temperature, and the more salt is needed in the brine.* **8.** Slowly mix the ice cream with the paddle for 20 to 30 minutes, until the ice cream becomes thick and creamy. *Mixing serves to cool the ice cream evenly and to break large ice crystals into smaller pieces, making the ice cream smooth.* **9.** Scoop the ice cream out of the cannister and place it in the freezer in a covered container for at least 1 hour before serving it.

Variations:
1. Stir chopped walnuts into the pumpkin ice cream after it is partially frozen. **2.** Substitute 3 cups of peach or apricot purée for the pumpkin purée and 2 vanilla beans for the cinnamon and nutmeg.

14 *Fat & Oil*

❑ Plants and animals produce fats to store energy efficiently. Animals store fat throughout their bodies for times when food is scarce; plants store fat in their seeds to provide the fuel needed for sprouting. The fats found in different animals and plants vary considerably. Understanding the general structure of fat molecules and the properties of different types of fat, helps us to choose the best fat for any recipe.

❑ At room temperature, the fats from animals are usually solid, while those from plants are usually liquid. Fats that are liquid at room temperature are usually called oils. Common cooking oils include those from soybeans, corn kernels, sunflower seeds, olives, coconuts, safflowers, cottonseeds, palm kernels, and peanuts. Animal fats used in cooking include lard (from pork) and tallow (from beef). Vegetable oil that has been modified chemically to make it solid is called shortening.

❑ All fats are composed of two building blocks: glycerol and fatty acids. The structure of a glycerol molecule is like the head of a fork with three short prongs. Fatty acids, on the other hand, are long chains composed of 4 to 24 carbon atoms. Fats are created when three fatty acids attach themselves to the three prongs of one glycerol molecule. These fatty acids can be identical or different. The result is a new, forklike structure with very long prongs. The drawing on this page illustrates how three fatty acids react with each glycerol molecule to create a fat molecule.

❑ The properties of all fats are determined by the particular fatty acids that join the molecule. One can, therefore, predict the properties of any fat by knowing the characteristics of the individual fatty acids.

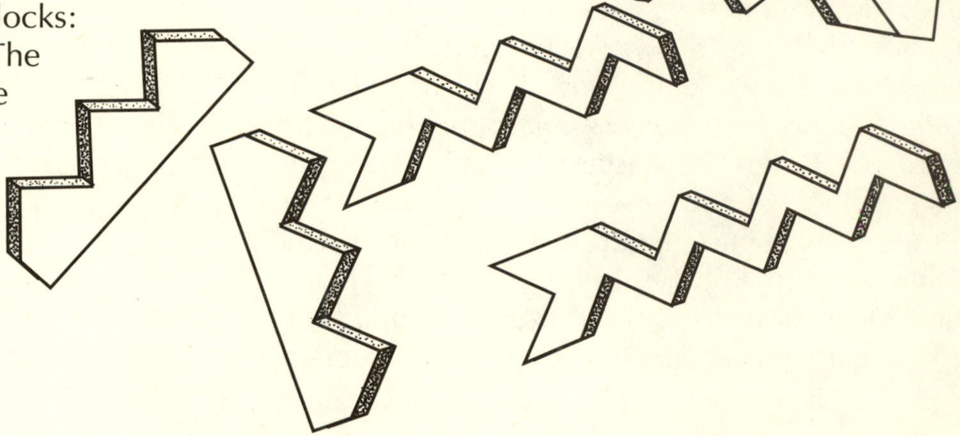

❏ The carbon atoms within a fatty acid chain must form exactly four bonds with other atoms. The bonds can be either single or double bonds, where a single bond uses one and a double bond uses two of the four available bonding sites. Generally, double bonds are stronger than single bonds.

❏ Carbon atoms linked together with single bonds have two free bonding sites. Hydrogen atoms can attach to these sites. Carbon chains that include double bonds have fewer free bonding sites available for hydrogen atoms to attach. Thus, carbon chains linked with only single bonds have more hydrogen atoms attached to the chain than do carbon chains that contain double bonds.

❏ Because carbon chains that contain exclusively single bonds have the maximum number of hydrogen atoms attached to the chain, they are consid-

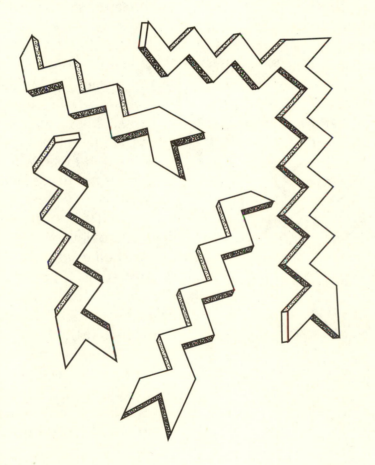

ered "saturated." Unsaturated fats, on the other hand, have at least one double bond in the fatty acid chain. This means that there are fewer hydrogen atoms attached to the carbon chain. (The carbon chain is not saturated with hydrogen atoms.) Fats that have more than one double bond are called polyunsaturated fats.

❏ Single bonds allow the carbon atoms to move freely, so that saturated carbon chains exist in a flexible zigzag. Double bonds, on the other hand, hold the carbon atoms tightly in one position and add inflexible wrinkles to the carbon chain. Unsaturated carbon chains are, therefore, kinked at each double bond. The drawing on this page shows several carbon chains. Three have only single bonds and the one on the right has a double bond, which causes it to kink in the middle.

❏ Fat molecules that can stack together are more likely to crystallize into a solid structure. Fats that do not stack together are more likely to be liquid. (See Chapter 6, "Sugar," for a similar situation.) Therefore, saturated fats, which are composed of long, regular chains that stack together easily, are usually solid at room temperature. Stacked saturated fats are illustrated on this page.

❏ Unsaturated fats have inflexible double bonds that force the molecule into a wrinkled configuration. Since fats with this configuration do not stack easily, unsaturated fats are less likely to form a solid

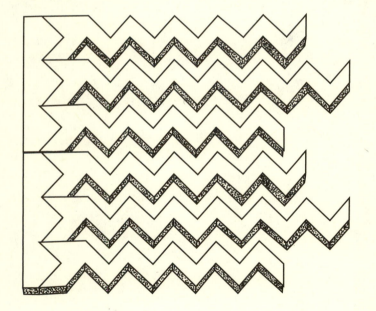

crystal and are usually liquid at room temperature. The illustration on the facing page shows a single unsaturated fat molecule.

❏ Most solid fats are composed of both saturated (crystalline) and unsaturated (liquid) fat molecules. The saturated fat crystals form a network that encloses the unsaturated liquid fat. Solid fats are pliable because of this combination of fat crystals and liquid fat.

❏ There is a direct relationship between the length of the fatty acid chains and the melting point of fats. Fats composed of fatty acids with long chains remain solid at higher temperatures because the long chains pack together more easily. This makes them more stable and less likely to melt when heated.

❏ If fats are heated to a very high temperature in the presence of water, they will smoke. The temperature at which this happens is called the smoke point. High heat provides enough energy to break down the fats into their original building blocks, glycerol and three fatty acids. Continued heating breaks the glycerol down further, producing a molecule called acrolein that evaporates and irritates the eyes. Smoking can be prevented by frying with fats that have high smoke points and by avoiding excessively high heat.

❏ Fats used in cooking should be appropriate for the specific recipe. In salad dressings, for example, the taste of the oil is most important, while the melting point and smoke point are irrelevant.

❏ In baking, fats are used to make cakes and muffins more tender. The fat coats and separates the flour, preventing excessive gluten formation, which would otherwise toughen the batter or dough. (See Chapter 16, "Yeast," for further discussion of gluten.) Oil is often preferred in quick breads such as muffins, because oil mixes easily with the other ingredients. On the other hand, hard fats, such as shortening and lard, are preferred when making biscuits and pie crust. These hard fats remain in distinct pieces and, when the dough is rolled out, form thin layers that separate thin layers of dough. The fat melts when the dough is heated but remains in thin layers, making the baked product flaky.

❏ Oils with a high smoke point are best for frying food quickly at a very high temperature in order to create a crispy outer crust. The cooking oil should be hot enough to keep the water in the food at its boiling temperature (212°F, 100°C). If the water is at its boiling temperature, the outward pressure of the escaping water vapor keeps the oil from soaking into the food. If the oil is not hot enough, it will seep into the food, making it greasy.

❏ Frying food in fat is faster than boiling food in water because cooking oil can be heated up to 450°F (232°C), while water can be heated only to 212°F (100°C). Frying is also faster than baking because the direct contact of the food with the hot oil transfers heat much more efficiently than does hot air.

❏ Fats used for frying also add flavor to the food. They react with the proteins and sugars on the surface of the food to produce the flavors and odors characteristic of frying. The high heat obtained with frying also promotes browning reactions on the surface of the food. (See Chapter 3, "Meat," for a more detailed discussion of browning reactions.) Using even a small amount of oil to prevent food from sticking in a pan produces these reactions.

❏ Most of the properties of fats and oils can be attributed to the number of double bonds within the fatty acid chains. These properties make different fats appropriate for different applications. In the diversity of nature, there is a perfect fat for every recipe.

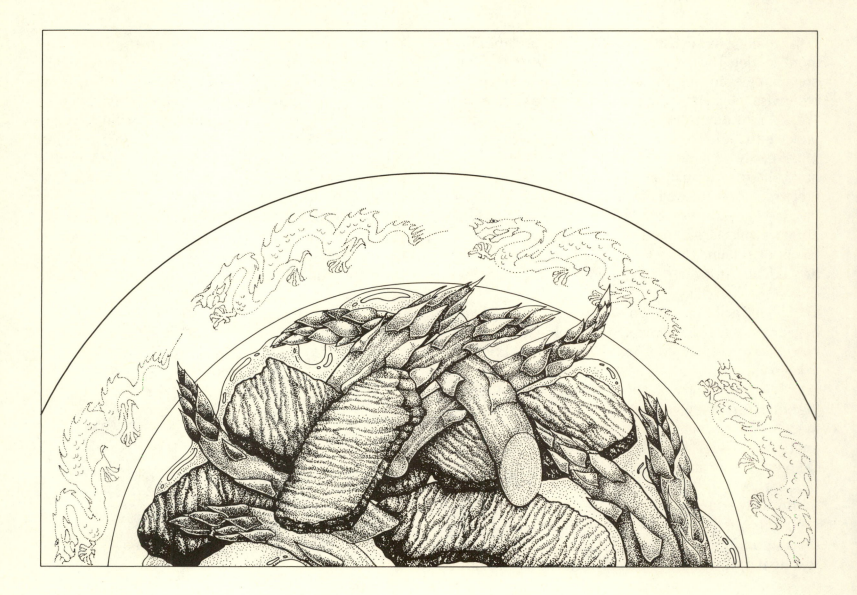

Chinese Beef with Asparagus

2 tablespoons oyster sauce
1 tablespoon soy sauce
1 tablespoon white wine
1 teaspoon sugar
A dash of black pepper
1 egg white
1 tablespoon cornstarch
1/2 pound flank steak, cut into thin slices
4 tablespoons vegetable oil
1 1/2 cups fresh asparagus, cut into 1-inch lengths
1/2 small onion, finely chopped
Ground red pepper, to taste

Preparation time: 1 hour
Yield: 2 servings

1. In a medium-size bowl, gently mix the oyster sauce, soy sauce, white wine, sugar, black pepper, egg white, and corn starch. *Dispersing the cornstarch in the cool liquids prevents lumps from forming.* 2. Add the steak and 1 tablespoon of oil, mix and set aside. 3. Steam the asparagus until it is bright green. 4. Heat 3 tablespoons of oil in a frying pan. Add the onion to the hot oil and sauté until soft and slightly brown. *Cooking the onion both drives off the pungent molecules and produces new sweet tasting molecules.* 5. Put the steak and sauce mixture and the sautéed onions in a wok. Cook, while stirring, until the meat is almost done, about 7 minutes. 6. Add the asparagus and cook until the steak is completely browned and the asparagus is heated through, about 3 minutes. Add the red pepper and serve immediately with white or brown rice.

Variations:
1. Substitute broccoli, green beans, or snow peas for the asparagus. 2. Add 1 or 2 cloves of crushed garlic. 3. Substitute boneless chicken breasts for the steak.

15 Oil & Water

❑ Oil and water do not mix. You can shake a container of oil and water for as long as you like, but as soon as you stop, the oil and water will separate. The water will sink to the bottom and the oil, because it is less dense, will float to the top. This property of oil and water can be applied to many areas of science, from cell biology to cooking.

❑ Molecules that dissolve freely in water ⟡ ⟡ are called hydrophilic (water loving) or lipophobic (oil fearing) while molecules that dissolve freely in oil are called lipophilic (oil loving) or hydrophobic (water fearing). Some substances, called emulsifiers ⟡—⟡, have both a hydrophilic head and a hydrophobic tail and, therefore, can interact with both oil and water, enabling them to mix.

❑ Soap is a common emulsifier. When soap dissolves in water, the hydrophilic head of each soap molecule is surrounded by water molecules. When the soap and water mixture comes in contact with an oily surface, the hydrophobic tails of each soap molecule are surrounded by oil molecules. To minimize the contact between the oil and water, the soap forms a bubble around the oil drop. The illustration on this page shows a soap bubble with the hydrophobic tails of the emulsifier molecules immersed in the bubble and the hydrophilic heads of the emulsifier molecules on the surface of the bubble. Because the outside of the bubble is hydrophilic, water can wash away the bubbles and the trapped dirt.

❑ Emulsifiers are used in cooking to facilitate the mixing of oil and water. Cooking emulsifiers do not actually allow the oil and water to mix freely, but, like soap, they allow small droplets of one to remain suspended inside the other. The final mixture is called an emulsion.

❑ A typical cooking emulsion is composed of water in which small oil droplets are suspended, with egg yolk, ground mustard, or complex sugars acting as the emulsifier. In addition to separating the oil and water, the emulsifier also helps prevent the droplets from pooling together. This happens because the hydrophilic head of the emulsifier has a small charge. Because the like charges on the neighboring emulsifier molecules repel one another, the droplets are less likely to pool together.

❑ To create an emulsion of oil in water, the oil must be separated into tiny droplets that are prevented from pooling together. Vigorous shaking, as seen in the drawing, breaks the oil into droplets. Emulsifiers are then used to keep the droplets from pooling together again.

❑ Emulsions can be temporary, like vinaigrette, or permanent, like mayonnaise. The delicate balance within an emulsion can be disrupted easily by exposure to extreme heat, cold, or shaking. The thicker an emulsion, the less likely it is to separate, because the droplets move more slowly through the thick mixture and are less likely to pool. This is one reason that mayonnaise does not separate easily and vinaigrette does.

❑ The ratio of oil to water determines the type of emulsion created: oil-in-water or water-in-oil. Because oil has a greater tendency to form droplets in water than does water in oil, an oil-in-water emulsion is most likely to form if equal parts of oil and water are mixed. With an increasing ratio of oil to water, there is a higher likelihood that the oil droplets will pool together. Finally, with a high enough oil to water ratio, the mixture will convert to a water-in-oil emulsion. For example,

vinaigrette made with two parts oil and one part vinegar (which is water based), with mustard as the emulsifier, will form an oil-in-water emulsion. If the oil-to-water ratio is increased to three to one, a water-in-oil emulsion will result.

❑ Some emulsifiers are better than others at keeping oil and water separated. Egg yolks, for example, are such good emulsifiers that they can be used to create a stable oil-in-water emulsion in a mixture that contains up to 75 percent oil. Egg yolks are used as the emulsifier in mayonnaise, which is made with three parts oil and one part vinegar.

❑ Fresh milk is another example of an emulsion of fat droplets in water. With time, however, the creamy fat separates from the water and floats to the top. To prevent this separation, milk is homogenized to break the fat globules into spheres that are too small to float to the surface. This stabilizes the emulsion.

❑ Emulsifiers are used in many types of food, including sauces, salad dressings, ice cream, and cakes. They allow the mixing of hydrophilic and hydrophobic substances that otherwise could not coexist.

Mustard Vinaigrette

3 tablespoons Dijon mustard
2 tablespoons lemon juice
5 tablespoons red wine vinegar (the highest quality
   available)
1/2 cup olive oil (the highest quality available)
1 clove garlic, crushed
Salt and pepper to taste

Preparation time: 5 minutes
Yield: approximately 1 cup

1. Mix the mustard and lemon juice together. Stir in the vinegar and blend well. *This helps keep the emulsion from separating.* **2**. Add the olive oil and mix well. *The vinegar will form tiny droplets in the oil and the mustard will serve as the emulsifier.* **3**. Add the garlic, salt, and pepper and mix well. Spoon the dressing over salad before serving.

Variations:
1. Substitute walnut oil or herb-flavored oil for the olive oil. **2**. Add 1 tablespoon fresh, chopped dill. **3**. Add 2 tablespoons of plain yogurt both to thicken the dressing and to act as an additional emulsifier.

Microbes

16 Yeast

❑ Yeast are very hardy, microscopic, single-cell organisms that can survive in a dry, inactive state for long periods of time. If inactive yeast are placed in a warm, wet solution and are given food in the form of sugar or starch they become active. The drawing on this page shows yeast cells growing and multiplying.

❑ In the presence of oxygen, yeast cells digest sugar or starch and release carbon dioxide gas bubbles and water as waste products. If there is little oxygen present, the yeast cannot fully digest the sugar. Under these conditions, they release carbon dioxide and alcohol as waste products. In cooking, we use both the carbon dioxide and the alcohol produced by yeast. When making wine or beer, the alcohol is collected and the carbon dioxide gas is allowed to evaporate. (See Chapter 17, "Wine," for details.) When making bread, the carbon dioxide bubbles are trapped in the dough, causing the bread to rise. The alcohol, meanwhile, evaporates.

The use of trapped gas bubbles to make dough rise is called leavening. When heat is applied during cooking, the bubbles expand within the dough and cause the dough to rise around them. The trapped gas bubbles make a cake or bread light and delicate. Besides yeast, baking soda (see Chapter 20), and baking powder (see Chapter 21) are also used to leaven batters. Each of these agents creates gas bubbles in a different way.

❑ Yeast bread requires several essential ingredients besides yeast, including flour, water, and salt. Flour forms the main body of most breads. When water is added to flour, a dough forms. The water binds to the proteins in the flour and produces a substance called gluten, which is a tangled network of large proteins and water. Gluten 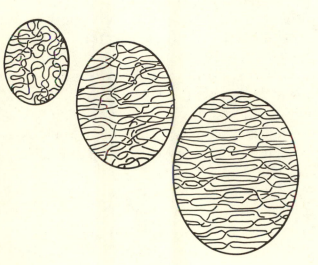 is very stretchy and retains the tiny gas bubbles produced by the yeast. As bread dough is kneaded, the gluten aligns. This alignment makes the bread dough smooth and silky. The drawing on this page illustrates how gluten aligns with kneading.

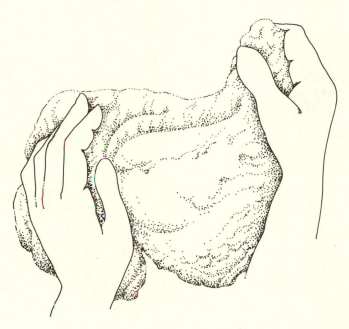

❏ Flour also contains starch ⋀⋀⋀ , which serves two major purposes in yeast bread. First, the yeast cells digest the starch by breaking it down into its constituent sugar molecules. Second, starch gives bread its firm structure. The starch combines with water in the dough and forms a gel, which helps to support the structure of the bread in much the same way that it thickens sauces. (See Chapter 7, "Starch," for details on how starch thickens sauces.)

❏ The salt ⋮ ∘• in yeast bread has two major functions: it adds flavor to the bread and, more important, it inhibits enzymes in the bread dough that break down gluten. If the gluten is broken down, the bread expands too rapidly and the holes produced by the expanding gas bubbles become too large.

❏ It is possible to make bread without salt by adding extra yeast. Extra yeast shortens the fermentation time necessary to produce the required volume of carbon dioxide gas. With a shorter fermentation, there is less time for the enzymes to break down the gluten.

❏ Salt also slows the activity of the yeast by drawing water ∘⋄ out of the cells by osmosis. Osmosis is the tendency of water to move through a membrane from an area of low concentration of dissolved particles to an area of higher concentration of dissolved particles.

❏ Yeast cells contain a solution with a low salt concentration. If the salt concentration outside the cells is increased above the concentration inside the cells, the water in the yeast cells flows out of the cells and into the surrounding solution. Therefore, if there is too much salt in bread dough, the yeast cells dry out and are unable to produce as much carbon dioxide gas. This prevents the bread from rising properly. The goal in making bread is to add just enough salt to inhibit the enzymes that break down the gluten, but not enough to dehydrate the yeast.

❏ Most bread recipes also call for a small amount of fat and sugar. Fat adds flavor and lubricates the gluten so that it slides and aligns more easily. Sugar sweetens the bread and is another source of food for the yeast. Too much fat or sugar, however, will slow gluten formation by inhibiting the interaction of the flour and water.

❏ After all these ingredients have been combined, they react with one another to produce bread dough. The yeast consume the sugar and the starch and release carbon dioxide gas bubbles into the dough. The water combines with the flour protein to form gluten; kneading the dough causes the gluten to align. The water also reacts with the starch and creates a gel. The salt adds flavor and helps to maintain the delicate gluten fibers. The alcohol that is also produced as a waste product of yeast digestion evaporates, filling the room with the sweet smell of fresh bread. The drawing on the facing page shows the structure of bread dough, including the large air bubbles, starch, and gluten.

❑ When the bread dough is finally put into a hot oven, the yeast cells are killed by the heat, stopping carbon dioxide production. The gas bubbles already in the dough continue to expand with the heat, making the bread rise even more.

❑ When the surface of the bread is dry and hard, the bread cannot expand any more. The surface of the loaf then begins to brown. Browning results when sugars and proteins in the dough react with one another, producing darkly colored and intensely flavored molecules. Browning reactions require very high temperatures and begin only when the surface of the bread is completely dry. Any moisture in the bread would keep the temperature at the boiling temperature of water ($212°F$, $100°C$). The browning reactions in bread are similar to those that occur in meat. (See Chapter 3, "Meat," for more information on browning.)

❑ After the bread browns, and the loaf sounds hollow when it is tapped, it is finally done. The best part is still to come... It's time to eat!

Braided Bread

1 cup milk
2 tablespoons honey
2 packages dry yeast
2 eggs, lightly beaten
2 teaspoons salt
4 tablespoons melted butter
4 to 6 cups unbleached white flour
Melted butter and sesame or poppy seeds for the top

Grease a large bowl
Grease and flour a baking sheet
Preparation time: 3³/₄ to 4 hours
Yield: 1 large loaf

1. In a small, heavy-bottomed saucepan, warm the milk until it is lukewarm, then pour it into a large bowl. Stir in the honey and then the yeast. Let the mixture sit for 3 to 5 minutes until it gets bubbly. *The yeast consume the honey and start to produce carbon dioxide gas bubbles and alcohol.* 2. Add the eggs, salt, butter, and 1 cup of flour and stir well. Add the remaining flour slowly, until it becomes too difficult to stir by hand. 3. Knead the dough on a floured surface, adding more flour a little at a time. When the ingredients are thoroughly mixed and the dough is the consistency of an earlobe, about 15 to 20 minutes, put the dough in the large greased bowl and cover it with a moistened towel. Let it rise in a warm place for 1 hour, until it has doubled in size. *The yeast will continue to consume the sugar in the dough and to produce the gas bubbles that make the dough rise.* 4. Slowly punch down the dough and knead it for 10 to 15 minutes, adding more flour if the dough is sticky. *Kneading helps to align the gluten molecules, making the dough smooth.* 5. Return the dough to the large greased bowl and again let it rise for 1 hour, until it has doubled in size. Punch it down and knead it briefly. Separate the dough into three balls of equal size. Roll the balls into long snakes and put them on the greased and floured baking pan. 6. Braid the dough on the pan. Cover the dough loosely with a moist towel and let it rise again for 30 minutes. 7. Preheat the oven to 350°F. Brush the risen dough with melted butter and sprinkle it with the seeds. Bake for 35 to 40 minutes, until the bread is golden brown and has a hollow sound when tapped. Let the bread cool before slicing it.

Variations:
1. Substitute molasses for the honey. 2. Use half white flour and half whole wheat flour. 3. Add 2 tablespoons cinnamon to the dough. 4. Substitute water for the milk. 5. Form the dough into any shape you desire.

17 Wine

❏ Wine is probably the most romanticized and the most analyzed food we eat. Despite a century of scientific research, however, winemaking is still an art.
❏ Wine is produced by fermenting grape juice with yeast 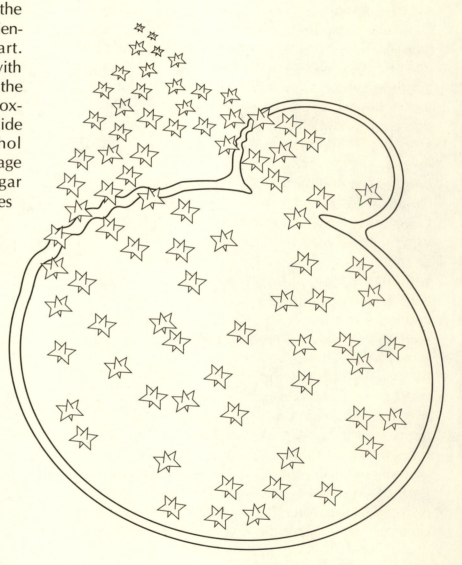. The yeast cells digest the sugar in the grape juice and release alcohol and carbon dioxide gas as waste products. The carbon dioxide gas escapes from the solution, and the alcohol turns the juice into wine. The drawing on this page shows a dividing yeast cell consuming sugar molecules. The drawing on the facing page illustrates how the yeast cells break down the sugar molecules into alcohol and carbon dioxide.
❏ Grapes are perfect for making wine for two reasons. First, grapes contain enough sugar (about 20 percent) to produce the desired amount of alcohol (about 12 percent). Second, grape juice is acidic enough to discourage the growth of many unwanted organisms that otherwise would spoil the wine's flavor. Although wine can be made from other fruits, flowers, and honey, these substance usually do not contain the proper balance of sugar and acid. The sugar and acid content must be adjusted in order to make wines from these other sources.

❏ Yeast, which are microscopic single cell organisms, are often found naturally on the skins of ripe grapes. The specific types of yeast used to make wine can live in the acidic environment of grape juice and can tolerate the relatively high levels of alcohol found in wine. The yeast used to make wine are the same as those used to make bread. However, when making bread, the carbon dioxide gas produced by the yeast is captured in the dough, causing it to rise, and the alcohol is allowed to evaporate. (See Chapter 16, "Yeast," for more information on using yeast to leaven bread.)

❏ Yeast can metabolize sugar with or without oxygen. In the presence of oxygen, yeast cells multiply very quickly and release carbon dioxide gas and water. Without oxygen, the yeast cells grow more slowly, because they do not fully digest the sugar, and release carbon dioxide gas and alcohol as waste products. Therefore, when making wine, the oxygen in contact with the grape juice is kept to a minimum. The alcohol produced by the yeast accumulates in the solution, and the carbon dioxide gas evaporates. Yeast cells also release small quantities of many aromatic compounds that enhance the wine's flavor and aroma.

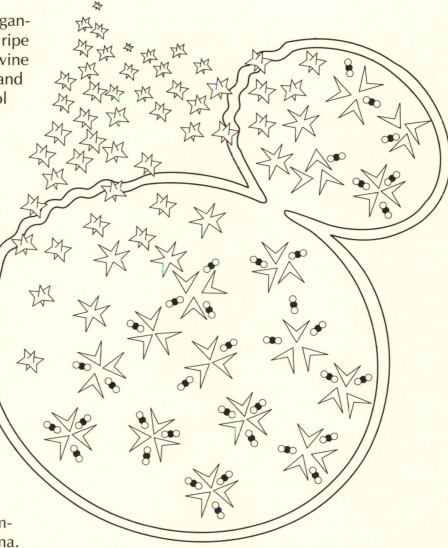

❑ Making wine is a complex process that starts with ripe grapes. There is an enormous variety of grapes from which to choose, each with a different color, size, and flavor, and each making a different type of wine. After the grapes have ripened on the vine, they are picked, taken to the winery, and crushed. The juice collected from the crushed grapes is called the must.

❑ The must is treated in several ways before yeast are added to the solution. Sulfur dioxide is added to the vat to kill unwanted microorganisms that would otherwise ruin the wine. Sulfur dioxide also prevents the must from browning by inhibiting the action of enzymes in the grape juice. (Enzymatic browning reactions are discussed in Chapter 12, "Tea," and in Chapter 22, "Lemons.") Sugar and acid may be added to the juice to ensure that the yeast cells have enough food and to make the wine tart enough.

❑ After these preparations are made, yeast are added to the must. The specific yeast strain is chosen for its rate of fermentation, alcohol yield, tolerance to alcohol, taste, and ease of removal from the wine at the end of fermentation.

❑ Adding yeast to the must begins the fermentation process. The vats are sealed and filled to the top to prevent exposure of the must to oxygen in the air, which would reduce the alcohol yield. A one-way valve on the top of the vat allows the carbon dioxide gas to escape from the vat. The drawing on the facing page shows two yeast cells that are consuming sugar and releasing alcohol and carbon dioxide gas.

❑ Fermentation continues for 4 to 6 weeks, until the yeast cells either consume all the sugar or die from the high alcohol content of the wine. When the alcohol content reaches 10 to 14 percent, fermentaion is complete. All the particulate matter, including the dead yeast cells, is allowed to settle to the bottom of the vat, and the wine is siphoned off. This process is repeated several times to clear the wine as much as possible. Sometimes gelatin, casein (milk proteins), or egg whites (albumin) are added to the wine in a process called fining. These proteins coagulate with particles floating in the wine and settle to the bottom of the vat, clearing the wine even more.

❑ After clearing, the wine is aged in large tanks or small barrels. Aging allows the many components of the wine to react with one another, causing the flavors to blend and meld. In some cases, the wine is aged in oak barrels that add an oaky flavor to the wine.

❑ When the initial aging is complete, the wine is transfered to bottles. The wine continues to age in the botttles until it is ready to drink. The bottles are stored on their sides to keep the corks covered with wine. This prevents the corks from drying out, shrinking, and allowing air into the bottle. Even a small amount of air can cause the wine to turn to vinegar. (See Chapter 18, "Vinegar.")

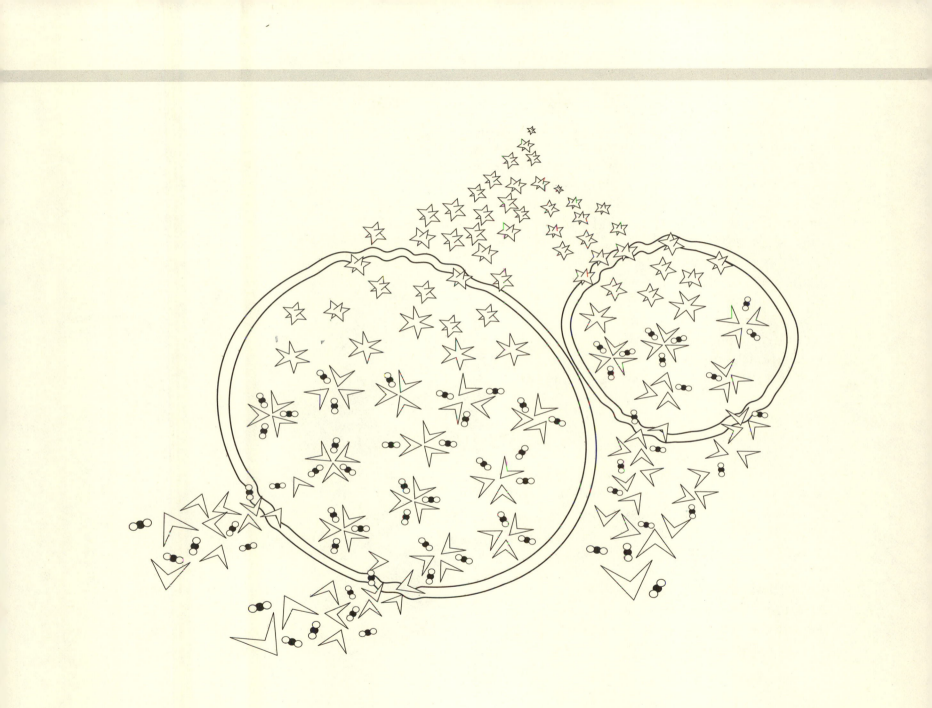

❏ Both the color of the grapes and the fermentation method determine the color of the wine. Wines made from light grapes are always white, whereas wines made from red grapes can be either red or white. To make white wine from red grapes, the clear juice is squeezed out of the grapes and the pigmented grape skins are discarded. Because the juice is clear, the resulting wine is white. Leaving the pigmented grape skins in the solution during fermentation produces red wine.

❏ Wines that do not contain much carbon dioxide gas are called still wines. Sparkling wines such as champagne, on the other hand, contain a considerable amount of dissolved carbon dioxide gas. To make sparkling wine, more yeast and sugar are added to a still wine. The yeast cells consume the added sugar and release carbon dioxide gas and alcohol as waste products, just as in the primary fermentation. Sparkling wines, however, are prepared in sealed containers so that the carbon dioxide gas produced by this secondary fermentation cannot escape. As a result, the gas remains dissolved in the wine.

❏ Sparkling wine can be made either in large vats or in individual bottles. In the latter case, the yeast sediment must be removed from the bottle in a complex two-step process. In the first step, called riddling, the bottles are twisted and turned over a period of a few months. After this time, the bottles are upside down with the sediment concentrated in the neck of the bottle. In the second step, called disgorgement, the sediment is frozen and the bottle cap is removed. The pressure of the dissolved carbon dioxide gas in the wine pops the frozen sediment plug out of the bottle. The sparkling wine is then topped off with more champagne and recorked. When a bottle of sparkling wine is opened, the pressure inside the bottle decreases quickly and the dissolved gas forms bubbles that rise to the surface of the wine and escape into the air.

❏ Winemaking is a complicated and time-consuming process. Nevertheless, wine lovers would agree that good wine is definitely worth the effort and the wait.

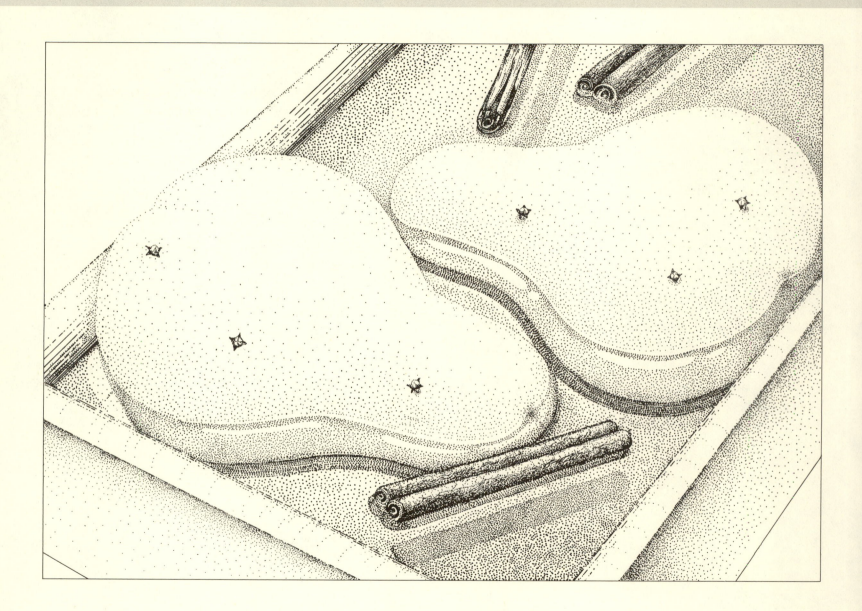

Wine Poached Pears

4 large, ripe pears, peeled and cored, cut in half
   lengthwise
Juice of 1 lemon
$\frac{1}{2}$ cup dry white wine
$\frac{1}{4}$ cup sugar
Whole cloves
3 cinnamon sticks
Ground cinnamon

Preheat oven to 350°F
Preparation time: 50 minutes
Yield: 6 to 8 servings

1. Put the pears, cut side down, in a glass baking dish
and sprinkle them with lemon juice.   **2**. In a separate
bowl, mix the wine and sugar. Pour the mixture over
the pears.   **3**. Stick 3 cloves into each pear half.
Place the cinnamon sticks in the liquid. Sprinkle the
pears lightly with ground cinnamon and cover the
baking dish loosely with foil.   **4**. Bake in a 350°F
preheated oven for 20 minutes. Baste the pears with
the wine sauce and bake for another 20 minutes.
Serve warm with vanilla ice cream.

Variations:
1. Sprinkle the pears with ginger instead of
cinnamon.   **2**. Substitute apples for the pears.   **3**. Use
red wine or dessert wine to vary the flavor.   **4**. Substitute
honey for the sugar.

18 Vinegar

❑ The word vinegar appropriately comes from the French words for sour wine. Vinegar is made from wine that has been exposed to both oxygen ◯ and to a specific bacteria, *Acetobacter* ◖◗. Sometimes this reaction occurs when it is not wanted, and perfectly good wine turns to vinegar.

❑ *Acetobacter* are usually present in the air and can thrive in an environment with high acidity and alcohol, such as wine. The bacteria digest the alcohol ⦨ in the wine and release acetic acid ⦨ as a waste product. Acetic acid gives vinegar its sour taste and aroma. The drawing on this page shows a sealed bottle of wine surrounded by air containing both oxygen and *Acetobacter*.

❑ Because the *Acetobacter* require oxygen to convert alcohol into acetic acid, winemakers can prevent *Acetobacter* growth by keeping air away from the wine. This is done by filling the wine vats to the very top with wine and by covering the vats. Wine can also be treated with sulfur dioxide, which kills

Acetobacter. If the wine-maker's goal is to convert the wine into vinegar, the wine is inoculated with the bacteria and the mixture is stirred to expose it to oxygen. In many cases, inoculation would not be necessary, because *Acetobacter* are so prevalent that they would probably find the wine anyway.

❑ Bottled wine should be stored in a way that prevents air and bacteria from entering the bottle to prevent *Acetobacter* from entering the wine. This is why wine bottles are often stored on their sides. In this way, the cork remains moist and does not allow air into the bottle. Once a bottle of wine has been opened, however, it is difficult to prevent the inevitable transformation of the wine into vinegar. Some people go to the trouble of filling the top of an opened bottle of wine with nitrogen gas to minimize exposure of the wine to oxygen.

❑ Vinegar can also be made from many other types of alcoholic beverages such as fermented apple cider, rice wine, or distilled alcohol. The process is the same in all cases: *Acetobacter* are introduced into the alcoholic beverage, and the mixture is stirred to incorporate air.

❑ The *Acetobacter* present in an alcoholic beverage consume the alcohol and release acetic acid. A thin film, called mother of vinegar, forms on the surface of the liquid after several weeks of exposure to air and *Acetobacter*. This film is a layer of *Acetobacter* that are multiplying on the surface of the liquid. The mother of vinegar can be transferred to another container to be used as a starter for more vinegar. The drawing on this page shows two opened bottles of wine. In the right-hand bottle, the *Acetobacter* have multiplied on the surface of the wine, formed a layer of mother of vinegar, and converted the alcohol into acetic acid.

❏ Most types of vinegar contain between four and five percent acetic acid and therefore have a similar acidity. However, different starting beverages produce vinegars with different tastes, colors, and aromas. The flavor can also be modified by adding other ingredients such as herbs, fruits, or garlic.

❏ Vegetables such as cucumbers, tomatoes, and onions are often pickled in a vinegar solution with sugar and spices. The acidic environment created by the vinegar inhibits the growth of microorganisms and thus preserves the vegetables.

❏ Vinegar is also added to most marinades to help tenderize meat. When the acidic vinegar contacts the proteins on the meat (including collagen) it denatures them. Because denatured proteins are softer, the meat becomes more tender. (See Chapter 3, "Meat," and Chapter 4, "Gelatin," for more information on denaturing proteins.)

❏ Finally, vinegar is used in salad dressings, where it is usually mixed with oil and spices. Because oil and vinegar do not mix, emulsifiers such as mustard or egg yolk are often added to stabilize the dressing. See Chapter 15, "Oil and Water," for more information on emulsions.)

❏ The transformation of wine into vinegar is similar to the conversion of milk into yogurt by lactic acid bacteria (see Chapter 19, "Yogurt"), and the conversion of grape juice into wine by yeast (see Chapter 17, "Wine"). Sometimes these reactions are planned and sometimes they happen on their own. Take comfort in the thought that if your wine turns to vinegar, you can always make vinaigrette!

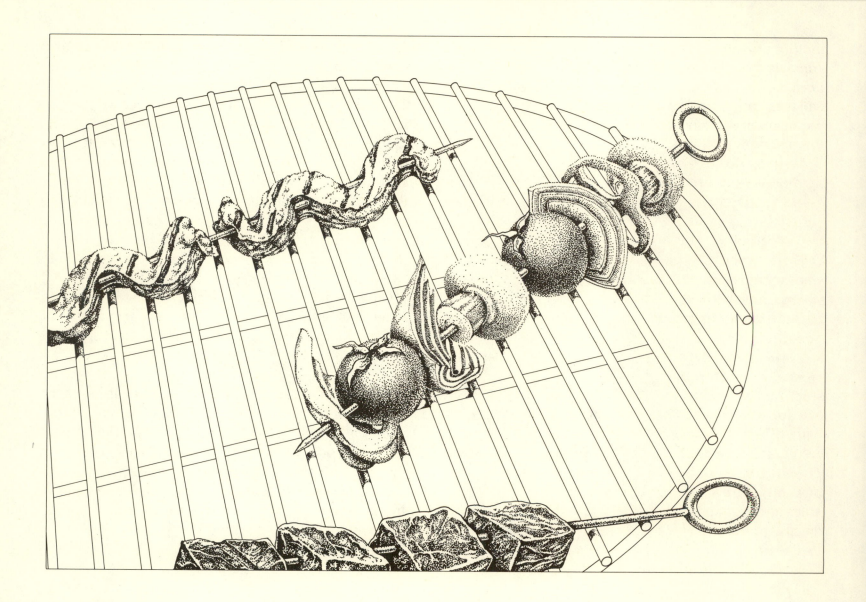

Shish Kebab

1/3 cup red wine vinegar
2 tablespoons Dijon mustard
1 cup olive oil
2 cloves garlic, peeled and crushed
1/4 teaspoon ground oregano
1/2 teaspoon dried basil
1/4 teaspoon dried thyme
Salt and pepper to taste
Approximately 1 1/4 pounds beef or chicken cutlets,
   cut into 2-inch cubes
1 whole red bell pepper, cut into 2-inch squares
1 whole green bell pepper, cut into 2-inch squares
4 medium onions, cut into wedges
10 large whole mushrooms
10 large cherry tomatoes

Preparation time: 2 to 3 hours, including marinating
Yield: 4 to 6 servings

1. In a medium-size bowl, mix the vinegar and mustard
and blend well. Add the olive oil and mix well. Stir in
the garlic, oregano, basil, thyme, salt, and pepper.
2. Put the beef or chicken cubes and the vegetables
in a large bowl and cover with the marinade.
Marinate the meat and vegetables in the refrigerator
for about 2 hours. *The acid in the vinegar tenderizes
the meat.*   3. Arrange the marinated meat and vegetables on skewers. Grill on a barbecue or broil until
the meat is cooked. Serve immediately with rice and
a salad.

Variations:
1. Add 1/2 cup red or white wine to the marinade.
2. Use any meat or vegetable combination you
choose. 3. Substitute 1 teaspoon ground ginger
and 3 tablespoons soy sauce for the oregano, basil,
and thyme.

19 Yogurt

❏ What happens when milk sours? If the conditions are right, it changes into a new food – yogurt.
❏ Yogurt is made by treating milk with two types of bacteria, *Lactobacillus bulgaricus* and *Streptococcus thermophilus*. Because these bacteria ⬭ consume milk sugar (lactose) and release lactic acid ⟿ as a waste product, they are called lactic acid bacteria. These bacteria are different from the bacteria that normally cause milk to spoil. The drawing at right shows lactic acid bacteria consuming lactose. The drawing on the facing page shows these bacteria releasing lactic acid as a waste product.

❏ Yogurt can be made from whole or skim milk. The milk is warmed to create an environment in which the bacteria can thrive, then a small quantity of lactic acid bacteria is added. These bacteria are found in commercial yogurt or may be obtained from a culture supply house. The mixture of milk and bacteria is then placed in a warm place for several hours to allow the bacteria to grow.

❏ As the bacteria grow, they convert the sweet lactose in the milk into lactic acid. This changes the sweet flavor of the milk to the sour flavor of yogurt. The presence of the lactic acid in milk increases the acidity of the mixture. High acidity causes the casein proteins ⊛ in the milk to bond together ∿∿∿, or co-agulate, forming curds.

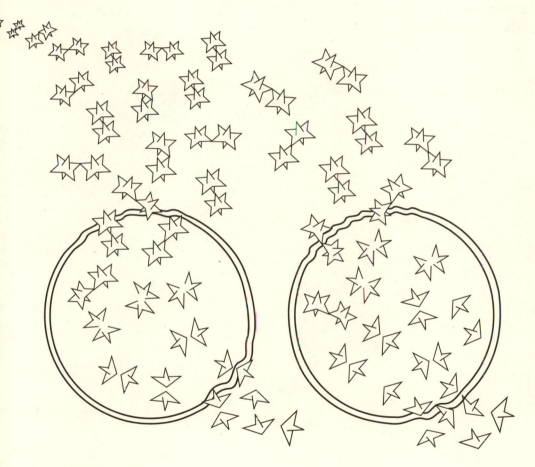

❑ Curds are essentially protein sponges that enclose a large quantity of the watery portion of the milk, which is called the whey. (See Chapter 2, "Milk," for more details about how proteins coagulate.) The drawing on this page shows curds forming in milk in the presence of lactic acid.

❑ The time that it takes for the bacteria to produce lactic acid and for the curds to form is determined by the temperature of the mixture. At higher temperatures the bacteria multiply more quickly, and the lactic acid accumulates faster. If the temperature is too high, however, the bacteria die and lactic acid production stops. Very low temperatures can also stop lactic acid formation by slowing the bacteria's metabolism.

❑ At 80°F (27°C) it takes about 17 hours for yogurt curds to form. If the temperature is increased to 100°F (38°C), the curds form in only 5 hours. Decreasing the temperature to 50°F (10°C) increases the time it takes to form curds to about 100 hours. If the yogurt is chilled in the refrigerator, curd formation essentially stops.

❑ Yogurt can be thickened by increasing the protein concentration of the milk because the higher the casein protein concentration in the yogurt, the more curds result. Adding powdered skim milk to the yogurt mixture is one way to increase the casein concentration. Another way is to boil the milk to evaporate some of the water.

❑ Yogurt is served in many different ways. Fruit or preserves may be added to make a sweet dessert or herbs and spices may be added for a savory dressing. Yogurt can also be added directly to batters. Because yogurt contains a very delicate balance of curds and whey, its structure can be disrupted by heat, salt, acidic ingredients, or vigorous stirring. Heating yogurt on the stove, for example, promotes bonding between the protein clusters and causes the curds to shrink. The water is squeezed out of the shrunken curds, just as water is squeezed out of the protein network in eggs when they are overcooked. (See Chapter 1, "Eggs.") Curdled yogurt does not taste unpleasant, but the small white curds floating in the watery whey do look odd. The drawing on this page shows what happens when yogurt curds shrink, forcing out the enclosed water.

❑ The curds and whey within yogurt also separate when yogurt is stirred and then returned to the refrigerator. When the yogurt container is reopened a day or two later, there is usually a watery layer on the top. This liquid is whey that has separated from the curds.

Yogurt Cucumber Dip

1¹/₂ cups plain yogurt
1 large cucumber, peeled, seeded, and finely chopped
3 tablespoons finely chopped red bell pepper
2 tablespoons minced onion
¹/₂ teaspoon cumin
Salt and pepper to taste

Preparation time: 40 minutes, including refrigeration
Yield: 8 servings

1. In a medium-size bowl, gently mix together all the ingredients. Refrigerate for 30 minutes.   **2**. Serve as a dipping sauce for pita bread and vegetables. *Use this dip soon after making it because stirred yogurt will separate into curds and whey, making it less appetizing.*

Variations:
1. Add a clove of minced garlic.   **2**. Add a tablespoon of chopped cilantro.   **3**. Use as a sauce over steamed vegetables.

Acids & Bases

20 Baking Soda

As discussed in Chapter 16, "Yeast," the development of gluten by kneading dough is an essential part of making yeast bread. Yeast produce carbon dioxide gas bubbles very slowly, and the rubbery gluten is needed to capture and hold the bubbles in the dough. However, because gluten ruins the texture of delicate cakes, muffins, and pancakes, the ingredients in these recipes are usually mixed only until they are moistened to prevent gluten formation. Yeast cannot be used to leaven these batters. In these cases, baking soda or baking powder, which produce carbon dioxide bubbles much more quickly than do yeast, are used as the leavening agents.

Baking soda, or sodium bicarbonate, is basic, which means that it has a pH greater than 7.0. When baking soda is mixed with an acidic liquid batter, the acid and base react quickly, neutralizing one another and producing carbon dioxide gas bubbles and water. The gas bubbles expand when the batter is placed in a hot oven, making the batter rise. The drawing shows the formation of the gas bubbles in the uncooked batter and the expansion of the bubbles in the hot oven.

The heat of cooking also promotes several other reactions that help to maintain the structure of the final baked product. First, the starch in the flour absorbs water and forms a gel, and second, the proteins in the eggs coagulate and form a protein network. (See Chapter 1, "Eggs," and Chapter 7, "Starch," for details.) The starch gel and the protein network create a structural framework for the expanding gas bubbles. This framework is maintained even after a cake, for example, cools and the gas bubbles dissipate.

Baking soda can be used to leaven a batter only if the batter contains acidic ingredients that can react with the basic baking soda. Common acidic ingredients include yogurt, sour cream, buttermilk, lemon juice, pineapple, vinegar, molasses, and honey. If the batter is not acidic, the unreacted baking soda will result in a bitter taste and the batter will not rise.

If too little baking soda is added to an acidic batter, the unneutralized acid remaining in the batter may change the flavor and texture of the final product. The acid may denature the egg proteins and break down the starch. As a result, the batter may not be able to contain the gas bubbles and will not rise properly.

It is very important to use the correct amount of baking soda in a recipe. Too much makes the food taste bitter and too little prevents the batter from rising. The correct amount of baking soda leavens the batter and is completely neutralized by acid.

Irish Soda Bread

2 cups all-purpose unbleached flour
1/2 teaspoon salt
3/4 teaspoon baking soda
1/4 cup sugar
2/3 cup raisins
2 teaspoons caraway seeds
2/3 cup buttermilk
1 egg
3 tablespoons melted butter or margarine
1 tablespoon butter or margarine for the top

Preheat oven to 350°F
Grease a round pie pan or flat baking sheet
Preparation time: 1 1/2 hours
Yield: 1 loaf

1. In a large bowl, mix the flour, salt, baking soda, and sugar. Stir in the raisins and caraway seeds. **2.** In a separate bowl, mix the buttermilk, egg, and melted butter or margarine. **3.** Pour the wet ingredients into the dry ingredients and mix only until moistened. *Mixing the wet and dry ingredients allows the baking soda to react with the acidic buttermilk to form carbon dioxide gas bubbles that, in turn, cause the dough to rise. Mixing the dough too much causes gluten to form, making the bread tough.* **4.** Put the dough in the greased pie pan or shape it into a circle and put it on the greased baking sheet. Dot the top with 1 tablespoon of butter or margarine. Cut an X on the top to protect the bread from cracking as it rises. **5.** Bake in a 350°F preheated oven for 50 minutes to 1 hour, until the bread is golden brown and sounds hollow when tapped. Serve warm.

21 Baking Powder

❏ Baking powder, which is essentially baking soda combined with an acid (such as cream of tartar), is used to leaven delicate batters that do not contain acidic ingredients. It is possible to make baking powder by combining one part baking soda with two parts cream of tartar. When water is added, the basic baking soda reacts with the acidic cream of tartar to produce carbon dioxide gas bubbles and water. The bubbles cause the batter to rise. Baking powder usually also contains starch, which absorbs moisture and inhibits the premature reaction of the baking soda and the acid.

❏ Baking powder reacts so quickly when it is placed in a liquid that most of the gas bubbles escape before the batter has cooked enough to trap the bubbles. The solution to this problem is double-acting baking powder.

❏ Double-acting baking powder has two different acids mixed with the baking soda. The first acid reacts with the baking soda at room temperature and forms small gas bubbles as soon as it is added to the liquid batter. The second acid reacts with the baking soda at higher temperatures and produces gas bubbles while the batter is in the oven. The second set of bubbles also increases the size of the small bubbles that were formed by the first reaction. The drawing on the facing page shows the formation of the first set of gas bubbles before the batter is placed in the oven and the formation of the second set of bubbles in the hot oven.

❏ By the time the second set of bubbles forms, the egg proteins within the batter have begun to coagulate and the starch has begun to gel. This makes the batter elastic enough to contain the bubbles. When the batter is fully cooked, all the leavening agents have been consumed, leaving behind only gas bubbles and water. As the cooked batter cools, the carbon dioxide gas bubbles dissipate, resulting in a light and delicate baked good.

❏ Some recipes call for both baking soda and baking powder. The batters in these recipes usually contain enough acid to react with a small amount of baking soda, but not enough acid to create the needed volume of carbon dioxide bubbles to leaven the batter thoroughly. Baking powder is added to supplement the action of the baking soda.

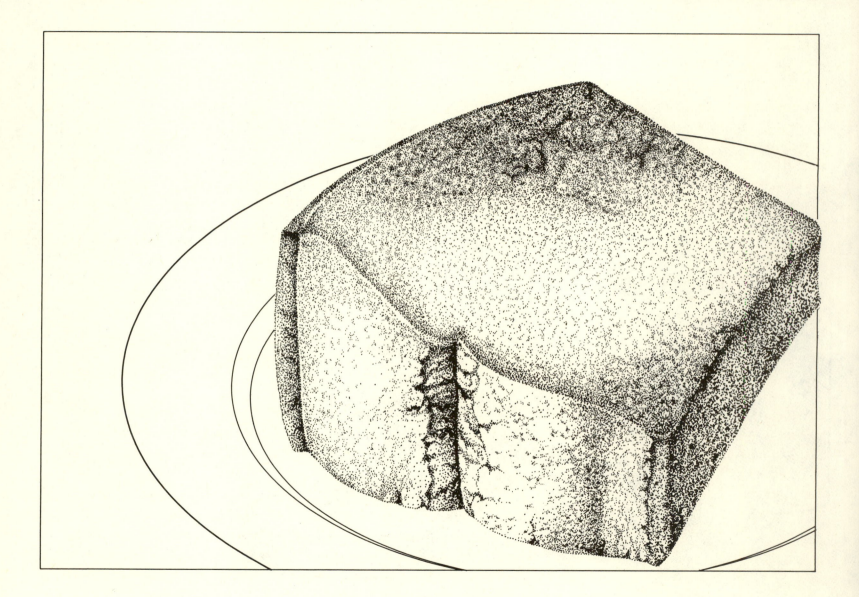

140

Pineapple Carrot Cake

3 large eggs
2 cups sugar
1 cup cooking oil
2 cups grated carrots (approximately 4 medium-size
    peeled carrots)
1 8-ounce can crushed pineapple
2 teaspoons vanilla extract
1 cup chopped walnuts
3 cups flour
1 teaspoon salt
1 teaspoon baking powder
1 teaspoon baking soda
1½ teaspoons ground cinnamon

Grease a tube pan
Preheat oven to 350°F
Preparation time: 1½ hours
Yield: 8 to 10 servings

1. In a large bowl, mix the eggs, sugar, and oil.   **2**. Stir in the grated carrots, pineapple (including the juice), vanilla, and walnuts.   **3**. In a separate bowl, combine the flour, salt, baking powder, baking soda, and cinnamon.   **4**. Slowly add the wet ingredients to the dry ingredients and mix just until moistened. *When the wet and dry ingredients are combined, the baking soda and baking powder each produce carbon dioxide bubbles that cause the batter to rise. Baking soda reacts with the moist acidic ingredients within the batter, whereas baking powder, which contains both baking soda and acid, produces bubbles when it becomes wet.*   **5**. Pour the batter into the greased tube pan. Bake in a 350°F oven for about 1 hour, or until a toothpick inserted into the middle of the cake comes out clean.   **6**. Cool the cake before removing it from the pan.   **7**. Serve alone or with vanilla ice cream.

22 Lemons

❏ Most of us have cut open beautiful apples, bananas, or avocados, then watched them turn brown before they are ready to serve. This browning process can be slowed by sprinkling lemon juice over the fruit. How does lemon juice slow browning?

❏ Fruit browns as a result of oxidation, which occurs when oxygen ◯ reacts with another molecule. Because oxygen has a high affinity for electrons, it steals them away when it comes in contact with a molecule that has loosely bound electrons. It, thereby, changes the properties of the donor molecule.

❏ Oxidation takes many forms and proceeds at dramatically different rates. The burning of wood, for example, is the result of rapid, uncontrolled oxidation. Our bodies, on the other hand, oxidize the food we eat in a slow, controlled way. Regardless of the speed of the reaction, oxidation releases energy stored in the bonds of the fuel molecules.

❏ Some fruits, such as apples and bananas, contain pigment molecules ∿∿ that react with the oxygen in the air when the fruit is cut open. The new molecules formed by the oxidation reaction are brown. Oxidation proceeds rapidly in some fruits because there are enzymes inside the cells of these fruit that help to promote the reaction of the pigments with oxygen. (See Chapter 5, "Onions and Garlic," for a description of enzymatic reactions.) The drawing below shows how the pigments within apples turn brown when exposed to oxygen.

❏ There are several ways to slow the oxidation of fruit pigments. If the reaction depends on enzymes, the enzymes can be inactivated by acids, salts, or the high temperatures of cooking. Each of these techniques denatures the enzymes, which

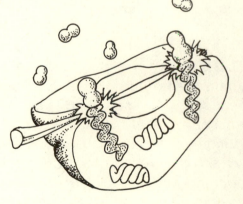

are proteins, and, therefore, decreases their ability to catalyze reactions. (See Chapter 1, "Eggs," for more information on protein denaturing.) Enzymatic reactions can also be inhibited by the low temperatures of refrigeration.

❏ Vitamin C 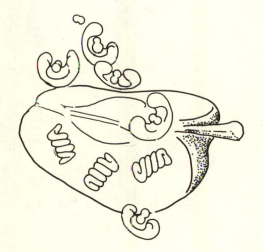, or ascorbic acid, also inhibits oxidation. An advantage of vitamin C is that it does not change the flavor, texture, or appearance of the fruit as much as does heat or the addition of salt.

❏ Many plants, including citrus fruits, pineapple, berries, green pepper, broccoli, and cabbage, synthesize vitamin C from glucose. (Vitamin C is also synthesized by many animals, though not by humans.) As a result of its unique structure, vitamin C reacts with oxygen so readily that it prevents oxygen from interacting with other molecules in the vicinity. It is essentially an oxygen trap. Because it interferes with the oxidation of other substances, vitamin C is called an anti-oxidant. As shown in the drawing on this page, vitamin C protects the pigment molecules in fruit from exposure to oxygen.

❏ Vitamin C is used commercially to slow the discoloration caused by the oxidation of various food pigments. You can do the same thing by dissolving a vitamin C tablet in water and then adding the mixture to fruits or vegetables that brown easily.

❏ Lemons contain high amounts of vitamin C and citric acid, both of which prevent the discoloration of fruit. Vitamin C acts as an antioxidant, and citric acid inhibits the enzymes that promote oxidation. Therefore, sprinkling lemon juice on fruits and vegetables that tend to turn brown prevents the delicate pigments from being destroyed by oxygen.

144

Summer Fruit Salad

2 peaches or nectarines
2 pears
2 bananas
1 cup of strawberries, cleaned with stems removed
1 cup of cherries, pitted
1 cup of grapes
1 cup sliced pineapple
The juice of 1 large lemon

Preparation time: 20 minutes
Yield: 6 servings

1. Slice the peaches, or nectarines, and the pears into bite-size pieces and place them in a large bowl. **2.** Slice the bananas and add them to the other fruit. **3.** Cut the strawberries in half lengthwise and place them in the bowl. **4.** Add the grapes and sliced pineapple. **5.** Sprinkle the fruit salad with the lemon juice. *The lemon juice prevents the pears and bananas from turning brown when exposed to oxygen in the air. The acid in the lemon juice directly inhibits the enzymes responsible for the browning and the vitamin C in the lemon juice acts as an anti-oxidant.* **6.** Serve the fruit salad alone or with yogurt, cottage cheese, or ice cream.

Variations:
1. Use any combination of seasonal fruit. **2.** Sprinkle the fruit with orange juice instead of lemon juice. **3.** Toss a few teaspoons of orange or almond liqueur over the fruit.

Glossary

acetic acid - The acid that forms when alcohol is oxidized. It is released as a waste product by *Acetobacter* and is responsible for the sour taste of vinegar.

Acetobacter - A type of bacteria that consumes ethanol (alcohol) and releases acetic acid as a waste product. It is used to make vinegar from alcoholic beverages.

acid - A substance that readily accepts an electron pair or donates a proton.

acidic - A property of a solution with a pH below 7.0. An acidic solution can be neutralized with a base.

acrolein - A molecule with a pungent odor that is created when glycerol is heated. It is produced when fats are heated to their smoke point.

albumins - Water-soluble proteins that are found in egg whites, blood, and milk.

alcohol - A molecule that has an -OH group (oxygen and hydrogen atoms) attached to a carbon chain.

alliin - A molecule found inside onion cells that reacts with alliinase, producing the strong odor characteristic of onions.

alliinase - An enzyme located between onion cells (in the intercellular space). It catalyzes the reaction involving alliin that is responsible for the odor of onions.

amino acid - A molecule that contains both an amino group ($-NH_2$) and a carboxyl group (-COOH). When amino acids bond together in a long chain, the resulting molecule is called a protein.

amylopectin - A long, branched starch molecule composed of glucose molecules.

amylose - A long, unbranched starch molecule composed of glucose molecules.

antioxidant - A substance that inhibits the oxidation of other substances. Vitamin C is an example.

ascorbic acid - A molecule, also known as vitamin C, that is found in fresh fruits and vegetable. It can function as an antioxidant.

astringency - A property of some drinks, such as tea and wine, that makes the mouth feel dry and tight.

atmospheric pressure - The pressure exerted on the earth and all objects on the earth by the weight of the air in the atmosphere.

atom - The smallest particle of any element that still has the properties of that element.

bacteria - Single-celled or multicelled organisms that multiply very quickly. They can live in air, soil, water, animals, plants, or rotting organic matter.

base - A substance that readily accepts a proton from an acid.

basic - A property of a solution with a pH above 7.0. A basic solution can be neutralized with an acid.

boil - The rapid conversion of a liquid to a gas with the eruption of bubbles. Boiling occurs when the vapor pressure of a liquid equals the atmospheric pressure.

bond - A chemical link between two atoms that arises when two atoms share electrons (covalent bond) or when two ions are attracted to one another (ionic bond).

bromelain - An enzyme that denatures proteins; it is found in pineapple.

caffeine - A molecule found in coffee and tea that stimulates many systems in the body.

carbohydrate - A group of molecules, including simple and complex sugars, that are found in plants and animals.

carbon - A common element that is found in many molecules, including sugars, fats, proteins, diamonds, and graphite.

carbon dioxide - A colorless gas produced by many reactions, including the fermentation of sugar. It is exhaled by animals and metabolized by plants.

caramelize - The process of breaking down sugar molecules into small fragments by heating it to a high temperature. The resulting mixture includes some dark-colored molecules.

casein proteins - A class of proteins found in milk that coagulate when exposed to acid, salt, or high temperatures.

catalyst - A substance that accelerates a chemical reaction but does not itself change as a result of the reaction.

cell - A microscopic biological unit containing intracellular fluid and a surrounding membrane. Many cells contain a nucleus.

cell membrane - A thin, sheetlike structure that forms the outside of a cell. It is composed of fats and proteins and separates the intracellular fluid from the extracellular fluid.

Centigrade (Celsius) - A temperature scale in which 0 degrees is the freezing point of water and 100 degrees is the boiling point of water. To convert to the Farenheit scale, multiply the Centigrade temperature by 1.8 and then add 32.

charge - Electrical energy of an atom or a molecule caused by an excess or deficiency of electrons. An excess of electrons results in a negative charge and a deficiency of electrons results in a positive charge.

chloride ion - A chlorine atom that has gained an electron and is negatively charged. It is present in table salt (sodium chloride).

citric acid - An acid found in many fruits, especially citrus fruits such as lemons and limes.

coagulation - The irreversible precipitation of proteins that have been exposed to acids, heat, salt, or alcohol.

collagen - Fibrous protein found in bones, tendons, and the skin of most animals.

concentration - The amount of one substance dissolved in another. It is usually measured as weight per volume, such as grams per liter.

condensation - The process of forming a liquid from a gas. Water, for example, condenses from steam.

covalent bond - A chemical bond in which two atoms share one or more pairs of electrons.

cream of tartar - An acidic molecule, also called tartaric acid, found in plants and fruits. It is added to baking soda to make baking powder.

crystal - A solid with a repetitive geometric shape. If broken, a crystal will maintain the same geometric pattern.

crystallization - The formation of crystals from a concentrated solution. Sugar crystals, for example, can form in a concentrated sugar solution.

curds - Coagulated casein proteins within milk.

cytoplasm - The part of a cell that is enclosed within the cell membrane. It does not include the nucleus.

denaturation - The process of destroying the original properties of a molecule by treating it with heat, acid, or salt.

dissolve - To disperse one substance in another; the dissolved substance is called the solute and the liquid into which it is dissolved is called the solvent.

double bond - A covalent bond in which two atoms share two pairs of electrons.

electron - A negatively charged subatomic particle that spins around the atomic nucleus.

element - A substance that cannot be broken down into other simpler substances without destroying the properties of the substance. Examples include oxygen, nitrogen, and carbon.

emulsifier - A substance that contains a hydrophilic and a hydrophobic end, allowing the formation of an emulsion.

emulsion - A mixture such as oil and water in which one is suspended in small droplets within the other.

energy - The capacity for doing work. There are several forms of energy, including heat energy, electrical energy, and mechanical energy.

enzyme - A protein that catalyzes a specific chemical reaction in a biological setting. An enzymes is not itself changed by the reaction.

evaporation - The conversion of a liquid into a gas, such as water into steam.

Farenheit - A temperature scale in which 32 degrees is the freezing point of water and 212 degrees is the boiling point of water. To convert to the Centigrade scale, subtract 32 and then multiply by 0.56.

fast-twitch fibers - Muscle cells, also known as white muscle fibers, that are very active for brief periods of time and use blood sugar as a source of energy.

fat - A molecule that is created when glycerol combines with three fatty acids. It is stored by animals and plants as an efficent source of energy.

fatty acid - A long carbon-containing molecule with an acid group on one end. Three fatty acids attach to one glycerol molecule to form a fat molecule.

fermentation - The process by which microorganisms break down sugars and release alcohol, lactic acid, acetic acid, or other subsances.

ficin - An enzyme found in figs that breaks down proteins and can be used to tenderize meat.

freeze - The conversion of a liquid into a solid at a specific temperature, the freezing point.

freezing-point depression - The decrease in the freezing point of a pure liquid when a solute such as salt is dissolved in it.

fructose - A simple sugar found in fruit.

galactose - A simple sugar that is formed along with glucose when lactose is broken down.

gas - A state of matter in which the molecules of a substance are far apart, move freely with a container, and fill any space available.

gel - A jellylike material containing up to 99 percent water with properties that resemble those of a solid.

gelatin - Denatured collagen that is used to create a gel.

glucose - A simple sugar that is the major source of energy for animals.

gluten - A complex, rubbery mixture of wheat proteins and water. It is formed when bread dough is kneaded and helps contain the carbon dioxide gas bubbles produced by yeast when making bread.

glycerol - A molecule containing three alcohol groups to which three fatty acids attach to form a fat.

hemoglobin - large, red-pigmented protein in blood whose major function is to bind oxygen in the lungs and to release oxygen to the tissues of the body.

homogenization - The process of creating a uniform solution. Milk is homogenized to break the fat globules into tiny spheres that do not separate from the water.

hydrophilic - The property of a substance that readily dissolves in water but not in fat.

hydrophobic - The property of a substance that readily dissolves in fat but not in water.

hygroscopic - The property of a substance that attracts or absorbs water molecules.

intercellular space - The space between the cells of an animal or plant.

ion - Any atom that has acquired a negative charge by gaining an electron or has acquired a positive charge by losing an electron.

ionic bond - A bond between oppositely charged ions that often breaks when a solid is dissolved in a liquid. Sodium chloride (table salt) is held together by ionic bonds.

lacrymator - A type of molecule produced by onions that causes tearing.

lactic acid bacteria - Bacteria that break down lactose (milk sugar) and release lactic acid as a waste product.

lactose - Milk sugar; it is composed of one glucose and one galactose molecule that are bonded together.

lipophilic - The property of a substance that readily dissolves in fats but not in water.

lipophobic - The property of a substance that readily dissolves in water but not in fat.

liquid - A state of matter in which the volume of a substance is fixed but the shape depends on the shape of the container. If a liquid is heated to its boiling point, it turns into a gas; if it is cooled to its freezing point, it turns into a solid.

melting - The conversion of a solid to a liquid at a specific temperature, the melting point.

melting point - The temperature at which a specific substance changes from a solid to a liquid.

metabolism - All of the biochemical reactions that are involved in breaking down molecules and building up new molecules in living cells.

microbe - An organism that is so small that it can be seen only under a microscope.

molecule - A distinct collection of atoms that are bonded together in a specific orientation. Examples include water, alcohol, fat, sugar, and starch.

mother of vinegar - A layer of multipying acetobacter growing on the surface of wine. They consume the alcohol in the wine and release acetic acid, turning the wine to vinegar.

muscle - A collection of elongated cells within an animal that can contract to produce movement.

must - The juice that is squeezed out of grapes or other fruit that is fermented to make wine.

myoglobin - A large protein found in muscle cells that binds oxygen from the blood for later use by the active muscle.

neutralize - To combine equal amounts of positively charged and negatively charged substances such that the charges are destroyed; often used with reference to acids and bases.

neutron - An uncharged, elementary particle located in the nucleus of an atom.

nucleus - In an atom, the nucleus contains the neutrons and protons; in a cell, the nucleus contains the DNA.

oil - A fat that is liquid at room temperature.

osmosis - The movement of a solvent, such as water, through a semipermeable membrane from an area of lower concentration of dissolved solids to an area of higher concentration of dissolved solids.

overrun - The amount of air added to ice cream, measured as the percentage that it increases the ice cream volume.

oxidation - The loss of one or more electrons from a molecule. Since oxygen readily accepts electrons, it often causes the oxidation of another substance.

papain - An enzyme that breaks down proteins; it is found in papayas.

pasteurization - The process of heating a substance, such as milk, to a high temperature to kill unwanted microbes.

pectic acid - A molecule formed from pectin in overripe fruit.

pectin - A carbohydrate molecule found in the cell walls of ripe fruit. Pectin is used to make preserves because it can form a gel in acidic solutions that contain a high sugar concentration.

pH - The negative logarithm of the hydrogen ion concentration of a solution. It is used to measure the acidity of a solution. A pH equal to 7.0 is neutral, a pH below 7.0 is acidic, and a pH above 7.0 is basic.

pigment - Specific molecules that give color to the cells of plants or animals.

polyunsaturated fat - A fat with more than one double bond in its fatty acid chains.

precipitation - The process by which a dissolved solid comes out of solution as crystals.

pressure - Force per unit area; e.g., pounds per square inch.

protein - A type of molecule that is composed of a chain of amino acids. Proteins fold into characterisic shapes and have many different roles within plants and animals.

proton - A positively charged elementary particle that is found in the nucleus of an atom.

protopectin - A type of carbohyrate molecule found in unripe fruit that is converted into pectin as the fruit ripens.

red muscle fibers - Muscle cells, also known as slow-twitch fibers, that contain a high proportion of myoglobin, are red in color, and are able to sustain long periods of activity in an animal.

riddling - The process of slowly twisting and turning a champagne bottle over a period of months in order to collect all the yeast sediment in the neck of the bottle.

salt - A crystalline solid that forms postively and negatively charged ions when dissolved in water.

saturated fats - Fats that have no double bonds. The carbon chains are saturated with hydrogen atoms.

seed - A crystal or particle of dust that initiates crystallization when introduced to a supersatured solution.

single bond - A covalent bond in which two atoms share a single pair of electrons.

sodium ion - A sodium atom that has lost an electon and is positively charged. It is present in table salt (sodium chloride).

solid - A state of matter in which the molecules are fixed in one place such that the substance has a definite shape. Heating a solid to its melting point causes it to become a liquid.

solution - A homogeneous mixture of one or more substances (the solute) that are dissolved in a another substance (the solvent.)

slow-twitch fibers - Muscle cells, also called red muscle fibers, that contain a high proportion of myoglobin, are red in color, and are able to sustain long periods of activity in an animal.

smoke point - The temperature at which a specific fat or oil will begin to smoke.

starch - A large molecule that is composed of chains of glucose molecules that are bonded together.

sucrose - A sugar molecule that is composed of one glucose and one fructose molecule that are bonded together.

sugar - A common name for any of various carbohydrates, including sucrose, fructose, and glucose.

supersaturation - The process of creating a solution that contains more solute than would normally dissolve at a given temperature.

tannin - A molecule found in many types of plants that binds readily to proteins. It has a bitter taste and astringent properties.

unsaturated fat - A fat that has at least one double bond in one of its three fatty acid chains.

vapor pressure - The pressure exerted by the gas above a liquid (or solid).

vitamin C - A vitamin, also known as ascorbic acid, found in many fruits and vegetables. It can act as an antioxidant.

volume - The amount of space a substance occupies, measured as cubic units, such as cubic feet or cubic meters.

whey - The watery portion of milk that remains when the casein proteins are coagulated by acid or salt.

white muscle fibers - Muscle cells, also known as fast-twitch fibers, that are very active for brief periods of time and use blood sugar as a source of energy.

yeast - A type of single-cell microorganism. Some kinds of yeast are used to leaven bread and to make wine because they consume sugar and release carbon dioxide gas bubbles and alcohol as waste products.

Index

acetic acid, *122 - 123*

Acetobacter, *122 - 123*

acid, *4, 6, 12, 13, 18, 24, 33, 53, 63, 64, 67, 78,114 - 116, 122 - 123, 125, 127, 129, 133 - 145*

acrolein, *98*

alcohol, *4, 108, 111, 113 - 116, 118, 122 - 123*

alliin, *36 - 38*

alliinase, *36 - 38*

amino acids, *3, 4, 10, 11, 25, 27, 30*

amylopectin, *50*

amylose, *50, 52*

antioxidant, *143, 145*

ascorbic acid, *143*

atmospheric pressure, *70 - 73*

atom, *1 - 4, 6, 96, 97*

bacteria, *19, 36, 64, 82, 122 - 123, 126 - 128*

baking powder, *109, 134, 138 - 139, 141*

baking soda, *109, 134 - 35, 137, 138, 141*

base, *6, 133 - 141*

boil, *19, 25, 26, 27, 46, 49, 62, 64, 70 - 73, 78, 85, 99*

boiling point, *26, 49, 72 - 73, 75, 99*

bond, *2 - 6, 11 - 13, 15, 18, 19, 25, 30 - 33, 44, 50, 52, 62, 97, 127, 129*

   covalent, *2*

bond (continued)

   double, *2, 3, 97 - 99*

   ionic, *2*

   single, *2, 3, 97*

bread, *44, 108 - 111, 113, 115, 134*

bromelain, *24, 33*

caffeine, *76, 79, 82, 85*

candy, *44 - 47*

   amorphous, *46*

   crystalline, *47*

caramelize, *64, 83*

carbohydrate, *4, 22, 27, 44 - 67*

carbon, *1, 3, 4, 96, 97*

carbon dioxide, *19, 108, 110 - 111, 113 - 116, 118 - 119, 134, 137, 138, 141*

casein, *17 - 19, 116, 127, 128*

catalyst, *4, 37, 143*

cell, *5, 6, 22 - 23, 25, 36, 38 - 39, 62, 64, 67, 108, 110, 114 - 116, 118, 142*

cell membrane, *5, 37 - 39, 64, 110*

charge, *1, 2, 6, 12, 17, 18, 62, 63, 73, 102*

chicken, *24, 29*

chloride ion, *2, 73*

citric acid, *143*

coagulation, *11 - 13, 18 - 19, 95, 116, 127, 134, 138*

coffee, *76 - 82, 85*
  decaffeination, *79*
  roasting, *76 - 77*
collagen, *3, 24 - 25, 30 - 33, 123*
cream of tartar, *12, 13, 46, 53, 138*
crystal, *6, 44 - 47, 49, 64, 88 - 93, 95, 98*
curds, *18, 127 - 129, 131*
cytoplasm, *5*
denature, *4, 6, 11, 13, 19, 24, 25, 27, 31, 33,*
  83, 95, 123, 143
deoxyribonucleic acid (DNA), *5*
dissolve, *5, 46, 73, 89 - 91, 102, 110, 119*
double boiler, *19, 21*
eggs, *10, 11, 12, 13, 33, 81, 95, 101 - 103,*
  113, 116, 123, 129, 134, 137, 138, 141
electron, *1, 2, 5, 142*
element, *1*
emulsifier, *16, 102 - 103, 105, 123*
emulsion, *102 - 103, 105*
energy, *3, 5, 22, 23, 45, 50, 96, 142*
enzyme, *4, 5, 19, 24, 33, 36 - 39, 64, 76,*
  82 - 84, 110, 116, 142 - 143, 145
evaporate, *19, 46, 65, 70, 73, 76 - 77, 108, 128*
fat, *10, 12, 13, 16, 17, 22 - 24, 53, 89, 93,*
  96 - 99,103, 111
  saturated, *24, 97 - 98*

fat (continued)
  unsaturated, *24, 97 - 98*
fatty acids, *96 - 99*
fermentation, *110, 114, 116, 118*
ficin, *24*
fining, *116*
fish, *24*
freeze, *88 - 92, 95*
freezing-point depression, *89 - 91*
fructose, *4, 44, 46*
fruit, *62 - 64*
frying, *26, 99*
fudge, *46 - 47, 49*
galactose, *19*
garlic, *36 - 39, 41, 55, 75, 105, 123, 125*
gel, *32 - 33, 35, 52, 62, 64, 67*
gelatin, *25, 30 - 33, 35, 62, 116*
glucose, *4, 19, 44, 46, 50, 143*
gluten, *98, 109 - 111, 113, 134, 137*
glycerol, *96, 98*
hemoglobin, *3, 22*
homogenize, *17, 93, 103*
hydrogen, *1, 3, 4, 97*
hydrophilic, *4, 102*
hydrophobic, *4, 102*
hygroscopic, *53, 63*

ice cream, *88 - 95, 103*

insulin, *3*

intercellular space, *36*

ion, *1, 2*

lacrymator, *38, 39*

lactic acid, *126 - 128*

lactic acid bacteria, *126, 127*

lactose, *16, 19, 126, 127*

leaven, *10, 109, 134 - 135, 138*

lemon juice, *12, 13, 53, 63, 64, 87, 105, 121, 134, 142 - 143, 145*

lipophilic, *102*

lipophobic, *102*

marinate, *24, 29*

mayonnaise, *35, 103*

meat, *22 - 27, 30, 33, 101, 125*

melt, *3, 45, 88 - 91, 98*

metabolism, *22, 23, 115*

microbe, *17, 108 - 131*

microwave, *26*

milk, *11 - 12, 16 - 19, 76, 85, 95, 103, 113, 123, 126 - 128*

mold, *65*

mother of vinegar, *123*

muscle fibers, *22 - 25, 27, 30, 79*
   fast-twitch, *23*

muscle fibers (continued)
   red muscle fibers, *22*
   slow-twitch, *22*
   white muscle fibers, *22, 23*

must, *116*

myoglobin, *22, 23, 27*

neutralize, *6, 12, 64, 134, 135*

neutron, *1*

nitrogen, *1, 3, 122*

nucleus, *1, 5*

oil, *76, 96, 98 - 103, 105, 125, 141*

onion, *35, 38 - 39, 41, 101, 123, 125, 131*

osmosis, *64, 110*

overrun, *92*

oxidize, *5, 23, 77, 142, 143, 145*

oxygen, *1, 3, 4, 5, 22, 23, 27, 77, 78, 83, 108, 115, 116, 122, 142 - 143, 145*

papain, *24*

pasteurize, *17*

pectic acid, *64*

pectin, *62 - 64, 67*

pH, *6, 33*

popcorn, *56 - 59, 61*

preserves, *44, 62 - 67*

pressure, *17, 57, 59, 61, 64, 70, 99*

pressure cooker, *72, 73*